Post Sound Design

The CineTech Guides to the Film Crafts

Series Editor: David Landau

Also available in this series:

Lighting for Cinematography by David Landau
Production Sound Mixing by John J. Murphy

Post Sound Design

THE ART AND CRAFT OF AUDIO POST PRODUCTION FOR THE MOVING IMAGE

JOHN AVARESE

BLOOMSBURY ACADEMIC
NEW YORK • LONDON • OXFORD • NEW DELHI • SYDNEY

BLOOMSBURY ACADEMIC
Bloomsbury Publishing Inc
1385 Broadway, New York, NY 10018, USA
50 Bedford Square, London, WC1B 3DP, UK
29 Earlsfort Terrace, Dublin 2, Ireland

BLOOMSBURY, BLOOMSBURY ACADEMIC and the Diana logo
are trademarks of Bloomsbury Publishing Plc

First published in the United States of America 2017
Reprinted 2018, 2019, 2020, 2021, 2023 (twice)

Cover design: Louise Dugdale
Cover image © Nebojsa Kuzmanovic / Getty Images

Library of Congress Cataloging-in-Publication Data
Names: Avarese, John, 1957- author.
Title: Post sound design : the art and craft of audio post production for the
moving image / John Avarese.
Description: New York, NY, USA ; London, UK : Bloomsbury Academic, an imprint
of Bloomsbury Publishing Inc., 2017. | Series: The CineTech guides to the
film crafts | Includes bibliographical references and index.
Identifiers: LCCN 2016052887 (print) | LCCN 2016055321 (ebook) |
ISBN 9781501327483 (hardcover : alk. paper) | ISBN 9781501327476
(pbk. : alk. paper) | ISBN 9781501307102 (ePub) | ISBN 9781501307126 (ePDF) |
ISBN 9781501327506 (ePub) | ISBN 9781501327490 (ePDF)
Subjects: LCSH: Sound–Recording and reproducing. | Motion pictures–Sound
effects. | Motion pictures–Editing. | Film soundtracks.
Classification: LCC TK7881.4 .A9628 2017 (print) | LCC TK7881.4 (ebook) |
DDC 621.389/32–dc23
LC record available at https://lccn.loc.gov/2016052887

ISBN: HB: 978-1-5013-2748-3
PB: 978-1-5013-2747-6
ePub: 978-1-5013-2750-6
ePDF: 978-1-5013-2749-0

Series: The CineTech Guides to the Film Crafts

Typeset by RefineCatch Limited, Bungay, Suffolk
Printed and bound in Great Britain

To find out more about our authors and books visit
www.bloomsbury.com and sign up for our newsletters.

CONTENTS

ACKNOWLEDGEMENTS

This book is a result of the two parts of my professional life that came crashing together. The first is that I have been a composer, sound designer, and film mixer for thirty years, working mostly in my home in Philadelphia. The second was becoming a college professor. I was not the most successful student in college, yet I managed to live a decent life doing what I love in a town that is not NY or LA. Once I found myself teaching a class, I was forced to explain what I had taught myself and what was instinctual for me. I was terrified. In time, and having the stubbornness to not give up, I was able to connect the dots in a presentable way to those students who were required to be in front of me. By teaching, I was learning why I worked the way I did.

I would like to thank Karin Kelly at Drexel University who gave me the opportunity to teach and learn in front of a classroom. It changed my life. I should also thank the filmmakers, whose shared experiences taught me the craft, and art, of storytelling. I must thank too Nick Natalicchio for helping me with graphics, and Joe DiVita for helping me put some of the concepts into words. And most importantly, I would like to thank my wife and daughter who have constantly supported me in my work, even when I disappear upstairs for days working with headphones on, unmindful of the world around me.

The information presented here is very much my way of working and thinking. Much like my teaching style, I am often telling stories about the topics discussed. You will find that I use “film” and “movie” interchangeably. That’s who I am. I am not wordy, and prefer to get to the point. I have worked with hundreds of producers and independent filmmakers. I love movies and I love making films. I hope this text will inspire you to do the same.

1 WHAT IS SOUND DESIGN?

Hearing is not like seeing. What is seen can be abolished by the eyelids, can be stopped by partitions or curtains, can be rendered immediately inaccessible by walls. What is heard knows neither eyelids, nor partitions, neither curtains, nor walls. Undelimitable, it is impossible to protect oneself from it. There is no acoustic viewpoint. There is no terrace, no window, no keep, no citadel, no panoramic lookout of sound. Sound rushes in. It violates. Hearing is the most archaic perception in the course of personal history, even before smelling, well before seeing, and it is allied with the night.

Pascal Quignard, *The Hatred of Music*[1]

WHY IS SOUND SO IMPORTANT?

Movies are like dreams. They send us to another reality. When we watch a really good film, we forget that we are watching something, but feel that we are *inside* it. I am sure this has happened to you. And then when the credits roll at the end, we are a little sad that is over and that we have to wake up to reality.

The role of a movie is to simulate reality. Most of how we perceive the world is through our sense of hearing: As we approach a street corner, our ears tell us that a car is approaching before we look at it. Bad audio pulls us out the film experience, however. When audio is bad, viewers think about the problems of the production instead of following the story, which is the opposite of what a film is supposed to do. Take for example, a scene that is shot in a large room, but where the focus is on an intimate exchange between a mother and a baby. The mother whispers tenderly to her newborn. We are drawn into the scene. However, if the audio sounds reverberate, the effect would be jarring, the tender moment is lost, and we wouldn't believe the story.

Sound also gives us a sense of space. A common example is a warm summer evening: A door from an interior opens and before we see the outside, we hear the sound of cicadas and insects informing us of the external setting. Sound can give us information that the picture or dialog cannot.

There are many layers of sound that surround us all of the time. Our perception of it is almost subconscious. For us to work in audio post production, we have to learn how to hear the layers independently. It is a very different way to listen. One student once told me in a post class that audio post production is everything that happens between the words. They were right.

WHAT IS SOUND DESIGN?

Every year, the Academy of Motion Picture Arts and Sciences gives two awards for sound: Sound Editing and Sound Mixing. Sound editing encompasses both all of the audio edited from production sound, and all of the sound that is added to the picture in post production. So what is included in this award are a number of individual disciplines. The dialog editor will edit and clean the production dialog. The automatic dialog replacement (ADR) recorder will record and substitute in any dialog lines that are needed after production has wrapped. The sound designers will create and edit and sound effects and ambiences (or atmospheres) in the movie. The person in charge of all of the sound editing is the supervising sound editor.

Sound mixing takes place once the sound editing is complete. All of these elements are then blended to make the final mix. In large studio action films, there may be up to 1,000 tracks of audio to use in the mixing decision process. The person who is charge of mixing the film is called the re-recording mixer. What he or she does is take all of the recordings from the sound editors/designers and create a new recording—in essence a re-record—for the final mix.

There probably could be six or seven Academy Awards for sound in films, but we'll take the two that we have.

LOW BUDGET PRODUCTION CONSIDERATIONS—AND A TRUE STORY

I have worked with many independent filmmakers as a composer, sound designer, and re-recording mixer all rolled into one convenient package. These directors hand over their baby to me to add an extra dimension, and give it emotion. They also need me to make the audio compliant to distributors' specs, and pass the all-important quality control procedure.

With low-/micro-/no-budget films, the effort has been put in the image. Directors can get decent cinematographers to shoot their project since it will show off their work. However, while directors control the set, audio is not a priority: Watching the images and concentrating on actors comes first, and typically little attention is paid to the production sound (it's assumed that the sound people will "worry" about the audio). The filmmaker will then go through the editorial process by editing it alone, or relying on a trusted colleague, who is often working on the project for free. Most of the time in these situations, they are listening on inadequate speakers that cannot accurately reproduce the audio recorded during production

After this process, they hand the film over to me and I will work on it by myself—usually handling a lot of damage control, like removing room noise, or editing out extraneous noises from the dialog. while making compromises in the audio—before the director, usually with their editor, comes in for the mix and we go through the film together. This is the first time that they *hear* their film in a proper environment with proper speakers, and gaining an education on film sound in the process. Many times this first session lasts about ten hours, and I can see the blood gradually drain from their faces. They often have no idea what was involved: In fact, on one project, the director told me that he knew he had bad audio, so he kept the sound volume as low as possible on his laptop computer, just barely enough to understand what was being said, so he did not have to confront the issue of bad sound.

IT ALL STARTS WITH CAPTURING THE SOUND

The above example isn't the only example of a filmmaker on a steep learning curve. In another incident, there was only one track of audio in a room where four people were having a dialog, and it did not sound good: It was very roomy. When I removed the reverberation from the audio, the sound became very "thin," or contained only high frequencies. This is uncomfortable to listen to, and the director asked why it sounded so bad.

I pointed to the screen. "Look at the ceiling," I said. Taped to the ceiling was a lavalier microphone; it was the only one used in the scene and was in the shot. (Nobody had noticed this in the editorial process, except me, of course.) It still had the mic clip attached to it.

The point that I am trying to make to filmmakers is that you must plan audio as meticulously as you plan your shots. While technology in audio post can do amazing things, it cannot fix source problems. In comparison, if the picture were out of focus, what could you do about it in post? Nothing.

So, as the person who is in charge of audio post production, whatever your title may be, meet with your director and producer *before* the shooting starts and implore them to get good, "clean" sound. The time, effort, consideration, and budget that they invest in good sound prior to shooting will pay off many times over when the project is in the final stages of post production and time is tight.

WHY YOUR LISTENING ENVIRONMENT IS SO IMPORTANT

Before you get started listening to your first project, it is important to consider how you are hearing. What system will be the reference that will determine your decisions? Every decision that you make will be based on your perception of how you hear the audio and every decision that you make driven by your opinion. Remember that it should not just be the speakers that affect your choices; the environment where you listen is an important consideration too.

Speaker Choices

Since you are attempting to recreate a movie experience in your audio post workspace, the closer you can get to a theater playback system, the better. While that may not be financially possible for many of us, I do not think that you have to spend a lot of money in order to obtain a system that will enable you to make important decisions confidently.

Choose speakers that are "full range," meaning that their frequency response is the same as the best of our hearing (20 Hz–20 kHz). Very few reasonably priced speakers extend that far. While the top higher frequencies may be reproduced, often the low end falls short of what is acceptable. If you can get speakers that reproduce low-end frequencies down to 40 Hz, you're in good shape. Reproducing frequencies below 40 Hz may require larger, heavier speakers with a large diameter woofer and a room that can handle the bass response.

Low-budget filmmakers are often surprised at the amount of low frequencies that exist when they get in a mixing environment. When I ask them what system they are listening to, they tell me that they listen on their laptop speakers. Typically, laptop computer speakers do not reproduce below 115 Hz. While I am aware that some film viewers may watch on their computers and smart phones, we, as audio post decisionmakers, must have the correct tools to assess our decisions.

You can easily find the frequency response of your existing speakers (monitors), or those you wish to invest in. Here are some examples of speaker choices and their frequency response, as of late 2016:

JBL LSR305 5″ Active Studio Monitor	43 Hz–24 kHz
Yamaha HS8 8″ Active Studio Monitor	38 Hz–30 kHz
Focal Alpha 80 8″ Active Studio Monitor	35 Hz–22 kHz
KRK ROKIT 10-3 G3 10″ 3-way Active Studio Monitor	31 Hz–20 kHz

Now you may think that a monitor with a frequency range that goes up to 30 kHz is better than those that cannot produce frequencies that high. But consider that our hearing, even at its best, can perceive only up to 20 kHz. So do not chose speakers based on numbers only. Go listen to a selection of speakers, and when do you so, select music that you know very well to use as reference audio. Use your ears and then choose the ones that you like.

If you already own a pair of speakers, you can listen and find out where they stop delivering low frequencies by playing sine waves of specific frequencies through your system. Most audio applications can generate a sine wave at any selected frequency. See what low frequencies your speakers are able to reproduce.

Speaker Placement

Now that you have selected full-range speakers (monitors) and can hear properly, you must carefully choose where to place them and, equally important, where to place yourself.

The science of acoustics and studio design is beyond the scope of this book, but there are a few simple guidelines to follow:

- Find the best possible place to sit. To do this, you have to minimize sound bouncing off flat surfaces, which act like a mirror. But instead of reflecting light, in this case they will reflect sound, so avoid rooms containing any glass-covered picture frames. Also try not to have your back to a wall, as sound will bounce and scatter directly behind you.

- If you are in a room that is not square, place yourself with the speakers located against the shorter width wall. This allows sound to travel the longest length of the room before it bounces. There is a variety of excellent room acoustic analysis software available if you want to find the best locations for your speakers, subwoofers, and listening positions.

Again, while this is outside the scope of this text, I do want to stress the importance of having the speakers on a horizontal plane with your ears. At a comfortable sitting position, the drivers of the speakers should be at the same height of your ears and be aimed directly at you. Also, your comfortable sitting position should be in the center of the two speakers. I cannot emphasize this enough. You are going to be making decisions about left–right panning, ambience that immerses the audience, and the level of dialog based on your listening environment. Your listening environment should have the same experience as wearing a pair of headphones. With headphones, one has a sense of a "phantom" center, where we experience sound that is coming from the center of our head; like it is in the middle of our forehead. Usually dialog is perceived from this phantom center. In music, the singer and some of the drums are usually in the center. In mixing for films, we are simulating reality where sound is located around us. For example, ambiences are mostly stereo, meaning that they should not be coming from the phantom center. If your speakers are set up correctly, when you listen to films, the dialog will appear to be coming from the center.

LISTEN

Now that you have configured your space and have some decent, full range speakers, it is time to listen to that room space as it's the one you'll be working in. So watch movies in that space, listen to music through those speakers. Since I am a one-person operation, I perform many tasks in my studio—from producing invoices to paying bills, maintaining equipment, etc—and there is always something playing in the system. I watch (and listen) intently to movies in my studio. I know my speakers and I know my room. By being familiar with my space, I can trust my decisions. This is probably the best recommendation that I can give anyone starting out in audio post production.

A Word About Headphones

I often get asked about working with headphones. I myself use them when I am going through a film for the first time to organize tracks or to make plans about what is going to be needed to tell the story to the audience. For me it is like working under a microscope, and headphones are good tools when performing surgery. ("Surgery" is the word I use to describe making miniscule edits, particularly with dialog.)

When it comes to balancing tracks, known as "mixing," I do not think that headphones are reliable, and here's why. When I teach a class in basic sound, I assign many audio projects that we listen to in class. By listening to hundreds of these, I have learned to pick out headphone mixes. Students are often very surprised at the unexpected difference in the way their mix sounds in a room. I cannot explain why headphone mixes have ambiences that are too loud when played against the dialog. I have prompted students on many occasions, thinking that the dialog is not loud because they know it so well and are not really focusing on an objective balance, but I do not believe that that is the case. In short, headphones are great for "surgery," but not for mixing.

STUFF TO REMEMBER

- Sound is a storytelling tool.
- We hear as much as we see, and often we hear before we see.
- It all starts with clean recordings from the shoot. Not everything is fixable in post.

PUTTING IT INTO PRACTICE

- Go outside and listen closely to all the sounds you are hearing. Which are louder? Which are softer? How do they overlap?
- Now listen to a movie or TV show and concentrate not on the dialog but all the sounds—the background, the sound effects, the music. Listen to how they "mix:" what gets louder and what gets softer, and when?

[1] Quignard, Pascal, Matthew Amos, and Fredrik Rönnbäck. "Second Treatise." *The Hatred of Music*. N.p.: n.p., n.d. 71. Print.

2 DAWS What's in the Box?

A Digital Audio Workstation (DAW) is a computer-based system for recording, editing, and mixing audio. Some DAWs are stand-alone, dedicated pieces of hardware, but most are based on computers that run audio-specific applications. Think of the stand-alone DAWS as a dedicated gaming device. Computer-based DAWs are superior because they are flexible for any specific application, and can also interface with third-party processors called *plugins*. As you might guess, the name for these processors stems from the fact that you can "plug' them into application. With the processing power of personal computers today, even an entry-level computer can handle mixing a short film and all of the elements described in this book.

Many video editors find that programs like Adobe Premiere or Avid Media Composer are good enough for editing audio, but DAWs are built for audio specifically and can therefore perform a wider range of functions than video editing applications. DAWs not only have a greater resolution for editing, but also include filters as well as noise reduction and automation processes that are built for sound. They can use third-party plugins that emulate vintage equipment and thereby make the history of audio hardware available to you. They can connect with audio hardware too. If you are thinking about making audio post-production a profession, investing in a DAW is a must.

There are many brands of DAWs, and the "industry standard" is the Pro Tools program. I started using it in the late 1980s when it was a two-track-only editing program called Sound Tools, and I have continued to use it since then as it has developed into the excellent program available today. In 1995 Avid bought Digidesign, who then owned Pro Tools, with the intent to create a seamless integration between video and audio editing. In this text, I will be referencing Pro Tools most of the time but other high-quality products are available: Adobe makes a wonderful program called Audition, which is part of their Creative Cloud suite and integrates with Premiere; Apple has Logic, another outstanding choice; many people use the affordable and flexible Reaper. I strongly believe that the best program for you is the one that you know the best. Whatever your choice, learn everything about the program. It's not about the tool that you own, it's about how you use it.

AUDIO INTERFACE

Eventually you are going to have to make a decision about purchasing an audio interface to connect to your computer. (An audio interface will be the way you connect a microphone to get sound into your computer.)

To understand how an audio interface records audio, and to help you make a decision about which audio interface is suitable for your needs, an explanation about the fundamentals of digital audio in necessary.

STRAIGHTFORWARD EXPLANATION OF THE FUNDAMENTALS OF DIGITAL AUDIO

When we make an audio recording, we are making an electrical representation of sound on a storage medium. We are not storing the actual sound, just as a photograph stores only a *representation* of a visual scene and not the actual subject.

What we hear as "sounds" are vibrating air molecules (compression and rarefaction). In order to be stored, this energy must be converted from its current form to an electrical signal via a *transducer*. In the case of turning sound energy into electrical energy, the transducer is a microphone. In the reverse process (that is, changing electrical energy back into sound energy), the transducer is a speaker.

Once the electrical signal has been converted, we can then transfer it to a storage medium. It is possible to store multiple signals at once, but this would require the use of two or more *channels*. An audio channel is an independent signal path for a given signal. People often—but incorrectly—use the terms "channel" and "track" interchangeably, but they are different entities. A track is a space on a storage medium; for example, the different songs on your iPod are tracks. The information that is transferred to your earbuds travels along two channels: One left, the other right. The following diagram shows this entire process.

Sound Energy Microphone (transducer) Electrical Energy Storage Medium

Analog vs digital

Presently we have two ways of storing this information: Analog or digital. For most of the twentieth century, analog was the only method of recording. It relies on making an analogy of the original form of sound energy (waveform) in its entirety. A copy is made using different processes depending on the storage media. For a piece of magnetic tape, iron particles are formed in patterns representing the information. For a vinyl record, grooves cut into the vinyl reproduce the waveform. When the needle runs over these grooves, it vibrates in the exact manner of the original sound waves. In an analog video recording, the shutter opens and closes to create an analogy of the light on the film.

In a digital recording, things are a bit different. Rather than creating an exact representation of the signal, the signal is translated into a series of zeroes and ones (binary code). Only this stream of information is actually stored. This is done by a process called *sampling*. In sampling, small sections of the waveform are taken (like little snapshots) at a very high rate of speed. When these snapshots are played back at the same rate of speed, it sounds continuous to our ears. Think of a cartoon, flip-book, or zoetrope—the static images go by so quickly that your brain is "tricked" into seeing movement. Here, the same concept is used for sound.

How fast or how often we take these samples is called the *sampling* or *sample rate*. Sampling rates are quite fast: 44,100 times per second or 48,000 times per second. One cycle per second is known as a hertz, while 1,000 cycles per second is a kilohertz. So, sample rates are always measured using units of kilohertz. It is important to distinguish sampling rates from frequencies of audible sound. Remember human hearing (at absolute best) is 20 Hz to 20 kHz. Sampling rates are always much higher numbers.

The sampling rate takes care of half of the information we need, however, as there are two sides to the entire equation of recording sound. On one side, we have frequencies, fundamentals, harmonics, and things of that nature; on the other, volume, dynamics, sound pressure level etc. These latter elements are represented through voltage, and the higher the voltage, the louder the sound. We measure this voltage using bits to give us a *sample depth* or *bit resolution*.

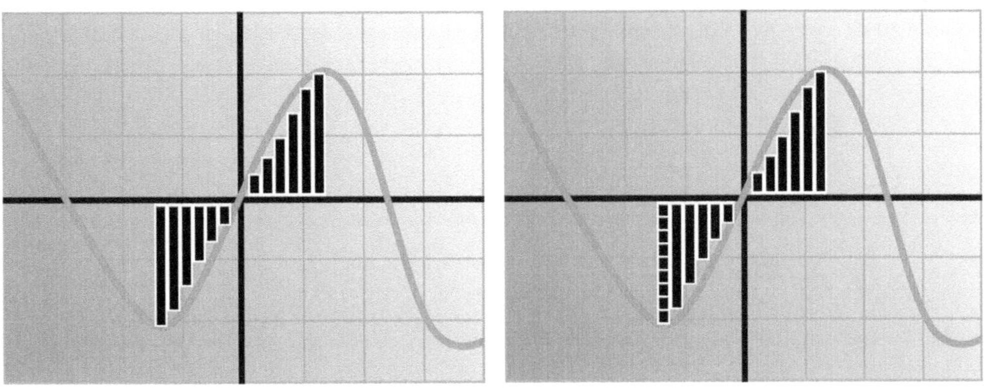

The recordings that you will be working with will be either 16 or 24 bits. Each bit can be either a one or a zero (on or off) and can store 6 decibels (dB) of dynamic range. These bits can be arranged in myriad combinations, giving us a total of 65,536 possible steps in 16 bit and a total of 96 possible dB of range, and 16,777,216 steps and 144dB for 24 bits.

Digital audio is represented on a bar graph: We measure sampling rate from left to right, and bit resolution from top to bottom. The processor that does all of this sampling is called an A to D (analog to digital) convertor.

This method of recording is not without its limitations, and one such is that in order for any given frequency to be stored and reproduced, the sampling rate must be at least twice that frequency. This theory was developed by electronic engineer Harry Nyquist, who was responsible for many technological advances in audio and communications in the mid-twentieth century. We refer to this concept as the *Nyquist Theorem* and the highest frequency that is able to be sampled (at any given rate) as the *Nyquist frequency.*

Another limitation is the size of the memory on the storage medium. These days memory is inexpensive and physically very small, so it is as much of a concern as it was in the early days of digital recording. Still, it is something to be mindful of. One can always calculate the size of a digital recording from the following rule: one minute of CD (compact disc) audio equals 10 MB of memory. Of course this relies on knowing the attributes of CD audio: CDs are always 44.1 kHz sampling rate at 16 bits and stereo (two independent channels).

In order to produce clean recordings, we must be careful about how loud they are. If recordings are made at too high a volume, sound will become distorted. There exists a range of volume information. The term we use to describe the ratio between the smallest and largest signals that can be measured is "dynamic range."

On the lowest side of the range is what is called the "noise floor"—the lowest level any signal can be captured at. The highest side of the range is "overload"—the point at which the signal distorts. For a 16 bit recording, the dynamic range is 96 dB, for 24-bit, 144 dB. If 0 dB represents the point of overload, -96 DB (144 for 24 bit) is the noise floor. A level of -12 dB is our "sweet spot" or reference level. All recorded dialogue should be at this level; anything above it is known as the *headroom*. Having adequate headroom is important as it gives the recording some room to "breathe" and space to put sounds that are louder than dialogue (think explosion sound effects).

Enemy number one in recording audio is distortion, which is defined as any change in the waveform from the original. The most common form of distortion is called *clipping* and is due to signal overload: The signal is too loud to be captured accurately and the peak of the waveform is cut off.

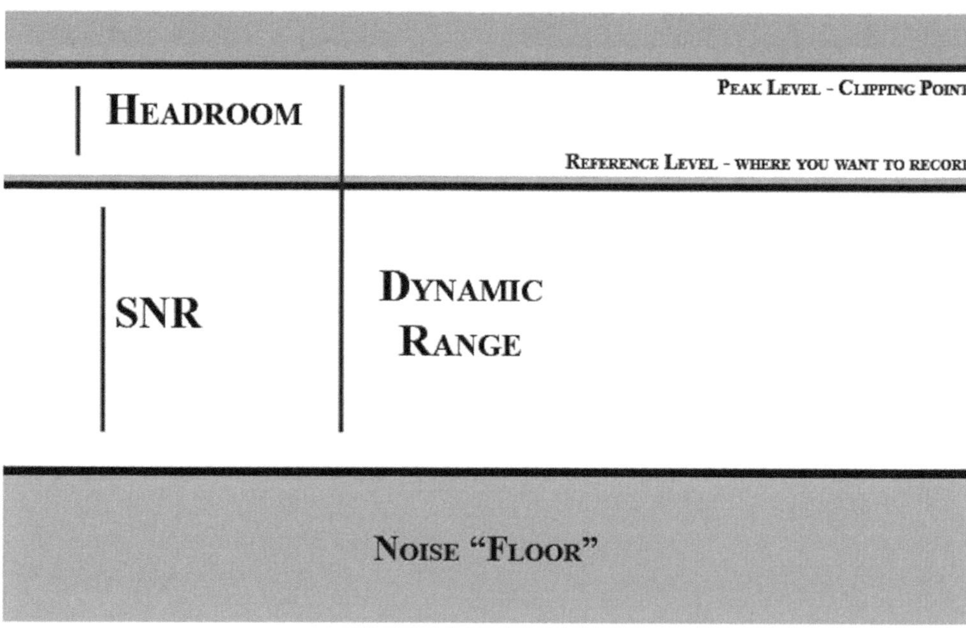

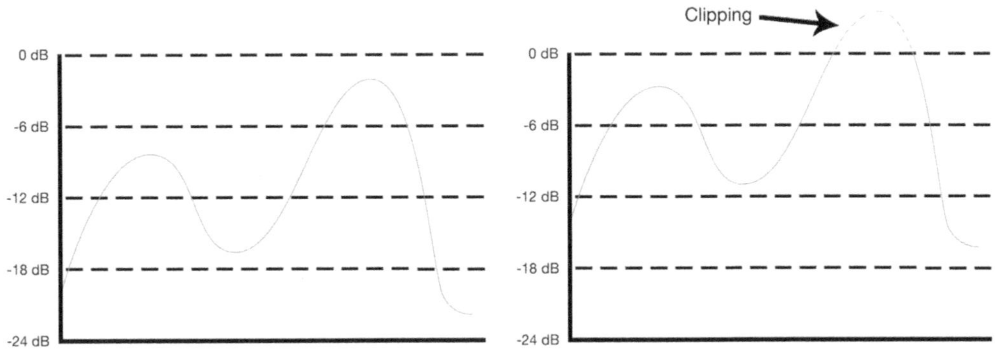

Clipping in an analog recording sounds distinctly different than clipping in a digital recording.

STUFF TO REMEMBER

- DAWs are specialized programs designed for high-resolution audio processing.
- Digital audio is a sampling of sound.
- Frequency response is determined by sample rate.
- Dynamic range is determined by bit depth.

PUTTING IT INTO PRACTICE

- Explore different DAWs in demo mode in order to get a clearer idea of their features.
- When watching a movie or TV show, actively listen for dynamic range.

---

3 AUDIO CONNECTORS, MICROPHONES

CONNECTING AUDIO EQUIPMENT

There are standard levels for audio channels, the lowest of which is known as *mic level*. A microphone output is at a very low volume, so it is connected to an input that boosts it to a higher volume known as *line level.* This is done with a device called a *pre-amplifier.* Sometimes pre-amplifiers are separate standalone units. Line level is different between professional and consumer audio units; the professional audio unit is louder.

Two switches are frequently found on microphones and on the inputs of preamplifiers. The first switch is called a *pad*, and it simply simply puts some "padding" on the input signal and lowers the volume by a determined value. This might be used when one is recording a loud sound effect and cannot get the input volume soft enough to prevent the signal from clipping. Pads are sometimes called attenuators; to "attenuate" means to reduce. The second switch is referred to as a *high-pass filter*, or *low roll-off*. This allows the higher frequencies on the spectrum to pass, while attenuating the lower frequencies. It is often used to filter out wind noise or any type of undesirable low frequencies.

Although not as common on a microphone, low-pass filters also exist, and in allowing lower frequencies to pass, remove the highs. The filter point is usually set at a fixed frequency.

Microphones are connected to their destination through cables. These cables are effectively antennas. They transmit signals, but because microphones transmit relatively weak signals and/or are susceptible to interference, we must use special wiring—namely *balanced* wiring with a specific *XLR connector.* Balanced cable has three wires inside: A signal wire; a ground wire; and a shield to reject hum caused by stray magnetic fields and interference from a variety of sources such as appliances, lighting, etc. In balanced lines, the shield/ground conductor does not carry any signal/reference itself, so can be left disconnected at one end to reduce "ground loops," or unwanted audible electric current. Unbalanced cables do not have a shield and are used to connect various other devices such as speakers, musical instruments, and signal processors.

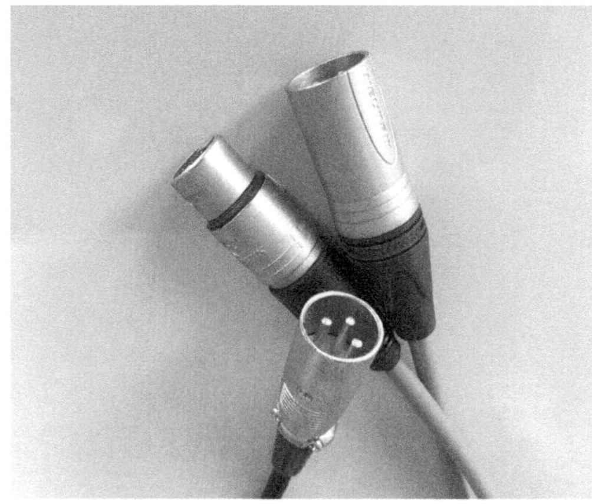

XLR microphone connectors

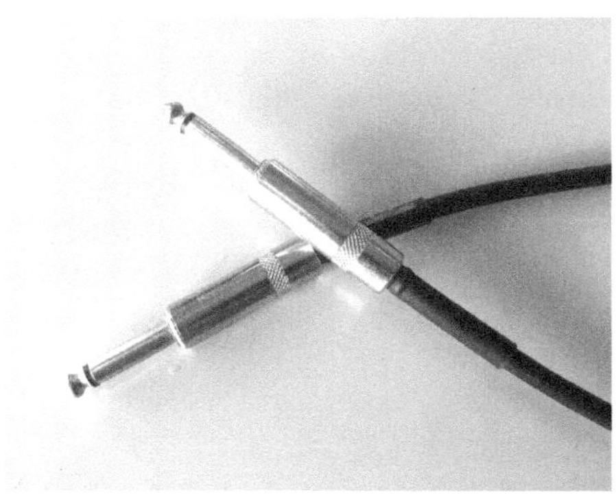

Unbalanced instrument patch cable, two conductors only

Headphone adapter

Headphone adapters are, of course, very widely used. They take a small headphone plug (one-eighth of an inch/3.5 mm) and convert it to a large one (one-quarter of an inch/6.35 mm), but this is an unbalanced connection, even though there are three conductors. Headphones are stereo: They have two independent channels and share the ground cable, so you have left signal, right signal, and one ground cable. No shield = no balance.

RCA connectors

Digital audio falls into two realms: Professional and consumer. Each uses different connectors and different wiring. Professional audio equipment typically uses higher voltage levels than consumer equipment, and it also measures audio on a different scale. Professional digital audio wiring as standardized by the Audio Engineering Society uses XLR connectors, balanced wiring, and has a longer cable distance before interference (due to the shielding). Consumer digital audio uses a standard called SPDIF (Sony/Phillips Digital Interconnect Format). It too features RCA connectors and unbalanced wiring, and works only over relatively short distances.

MICROPHONES

Understanding the fundamentals of microphones and their use is important as it allows us to effectively capture what audio we need and reject anything superfluous.

Suppose we wanted to record audio for a scene at a cocktail party. Imagine the sound of the cocktail party when you first enter the room. You hear lots of voices blending together in a nondescript blanket of noise. When you walk up to one individual conversation, that blanket of noise falls into the background and you can successfully distinguish individual voices. This can be accomplished in a recording only by using proper microphone technique with the correct type of microphone, or even more than one microphone.

But how do microphones work? They are basically mechanical versions of our ears. Waves of pressure enter the outer ear and beat against the ear drum or tympanic membrane. The vibrating of the ear drum moves three small bones which amplify the sound (hammer, anvil, and stirrup). There are some other interesting things in there like the "organ of corti" in the inner ear. All of these elements act as a biological transducer, changing the sound energy into electrical impulses that can be carried to the brain via the auditory nerve.

Microphones function in a similar way. The ear drum is replaced with a diaphragm. The diaphragm acts as a membrane and moves against a backplate, thereby changing the capacitance and creating an electric signal that is amplified (like the processes in the inner ear) and carried away via the cable (like the auditory nerve).

There are two main attributes or "categories" to any given microphone: The *type* of microphone (usually defined by the mechanics of how it works); and its *pickup pattern*. Typical types of microphone are: Dynamic; condenser; electret; ribbon. Pickup patterns are as follows: Omnidirectional; bi-directional or figure 8; cardioid; hypercardiod; and shotgun.

A microphone's pickup pattern describes how it "hears" sounds from different directions, and rejects sounds from the directions not required. Each pickup pattern has a specific use and creates a distinct sound by how it treats what it picks up.

Types of Pickup Pattern Explained

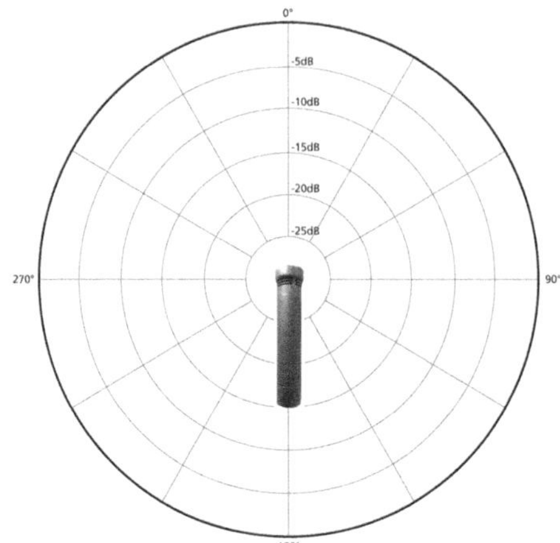

Omni pattern

- **Omnidirectional.** This pickup pattern receives sound from all directions equally. It is represented by a circle on diagrams. Although not particularly useful for recording dialog in film and television as it picks up a lot of room reverberation, it works well when recording music in interiors—such as orchestras—precisely because they do pick up on the room reverberation. In this latter case, the room (or concert hall) is the sound of the orchestra.

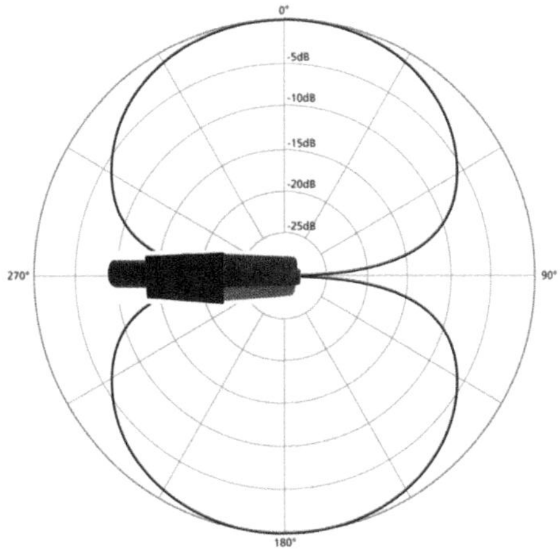

Bi-directional pattern

- **Bi-directional/Figure 8.** This pattern responds best to sound from the front and back while rejecting sound from the sides. Bi-directional microphones are used in recording and broadcast for their very neutral sound as well as in special stereo recording techniques such as mid/side (M/S), and special stereo microphones. All ribbon microphones are inherently bi-directional.

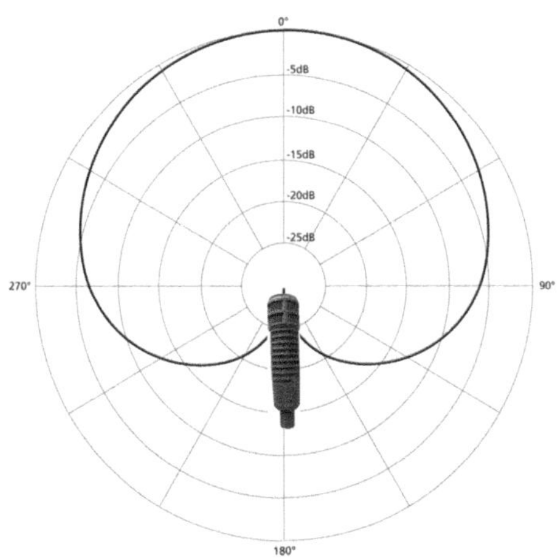

Cardiod pattern

- **Cardioid.** This uni-directional pickup pattern is shaped like a heart. In this case, an omnidirectional and a figure 8 (bi-directional) pickup pattern work together in harmony to create a single direction by putting signals out of phase. It is a popular pattern used for vocalists or voice actors, but is particularly susceptible to "proximity effect"; that is, the low, bass frequencies become accentuated at close range to the microphone.

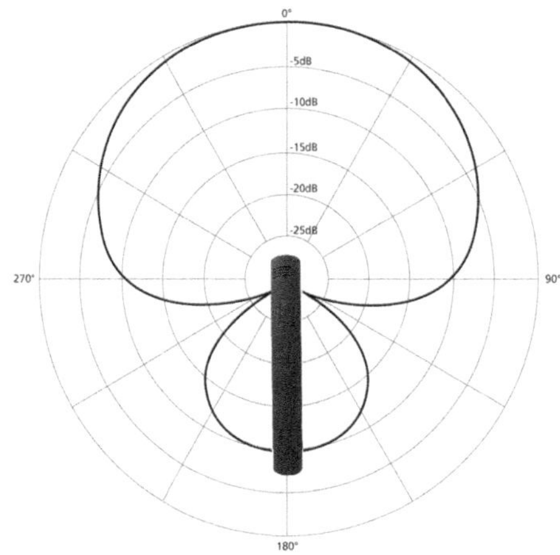

Hypercardiod pattern

- **Hypercardioid.** This pickup pattern is a narrower version of the cardioid, and has an even greater rejection of sounds coming from the sides. Because the front direction is more focused, a small lobe of sensitivity is created to the rear. This pattern is best for indoor dialog recordings and loud environments.

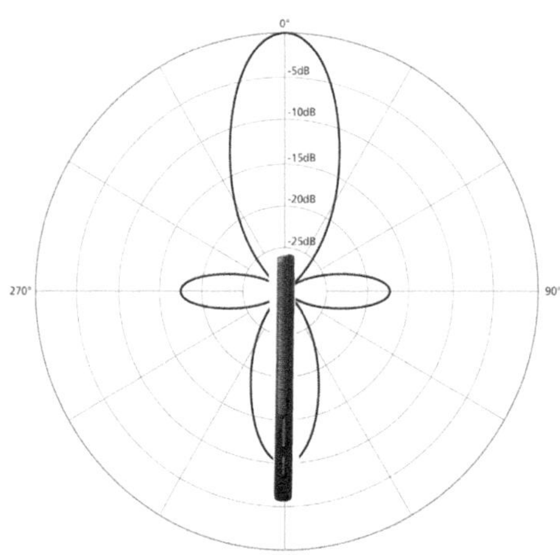

Supercardiod (shotgun) pattern

- **Shotgun.** This microphone pattern uses phase cancellation in a complex design to create an extremely narrow and focused pickup pattern. It is especially useful at long ranges. A small lobe also exists in the rear of the microphone. This is important to be aware of as it is possible to capture unwanted speech or reflected sound from that direction. They are excellent for use in film and theater to pick up sound while keeping the microphone out of the view of the camera. They are also frequently seen on the sidelines of sporting events. Longer shotgun microphones have narrower angles and more coloration and are best suited for exteriors and non-reverberating (dead/dry) rooms.

Microphone Types Explained

- **Dynamic.** Just like a speaker, a dynamic microphone contains a magnet that has a coil of wire around it that moves back and forth to create an electrical charge. Rugged and sturdy—they make great hammers—dynamic microphones are great for recording loud sounds as they are not as sensitive as other microphones, particularly at high frequencies. In addition, their sensitivity drops off at around two to three feet, meaning that they do not capture sound very well that is more than that distance away.
- **Condenser.** A condenser microphone is an excellent choice for capturing subtle sounds. It is a delicate, sensitive microphone that makes things sound very "alive," and works well on a boom pole. Condenser microphones have a moving metalized plastic diaphragm mounted close to a back plate. Perhaps the most important thing to remember about them is that because they lack a magnet, they must get their power from external sources, known as "phantom" power. It is usually 48 volts and supplied from either a battery or through the cable from a connected device.
- **Electret.** These microphones are technically a subcategory of condenser microphones. They are powered by static electricity. Electrets are small and thus particularly useful where space is limited. The most common use for an electret microphone is in a cell phone.
- **Ribbon.** A ribbon microphone works by suspending a small piece of aluminum foil—the ribbon—between two magnets. These microphones are extremely delicate, work on a different power scheme then the other microphone options, and produce the lower signal too. Ribbon microphones have a bi-directional pickup pattern, but some employ baffles on one side that can be engaged to block the sound, changing the pattern to uni-directional.

FREQUENCY RESPONSE

Any microphone, regardless of its type or pickup pattern, will have its own individual frequency response. That is, each microphone will be more or less sensitive to certain frequencies or bands (ranges) of frequencies. This is shown by plotting a line on a graph of the frequency spectrum that spans the audible range of human hearing (20 Hz– 20 kHz). This information is important for choosing a microphone or a specific application.

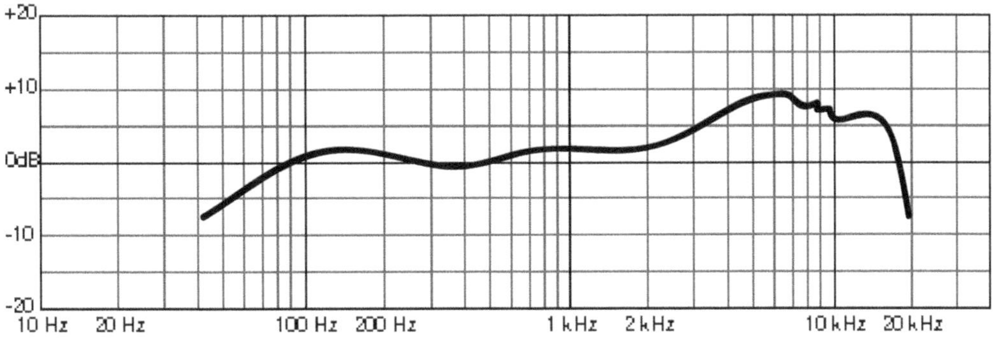

Dynamic microphone frequency response

With this dynamic microphone frequency plot, we can see that it does an excellent job of reproducing speech. As is typical with the range of speech, the low frequencies are almost non-existent and the high frequencies drop off around 12 kHz.

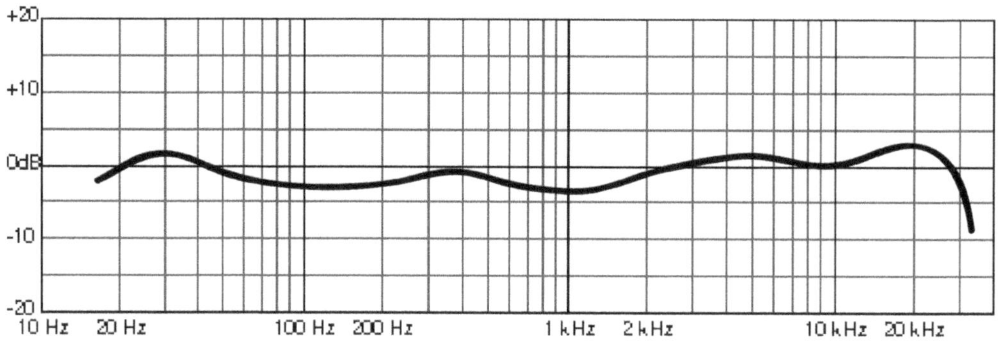

Condensor microphone frequency response

A condensor microphone has an almost flat frequency response throughout the range of our hearing and beyond, reproducing below 20 Hz and above 20 kHz.

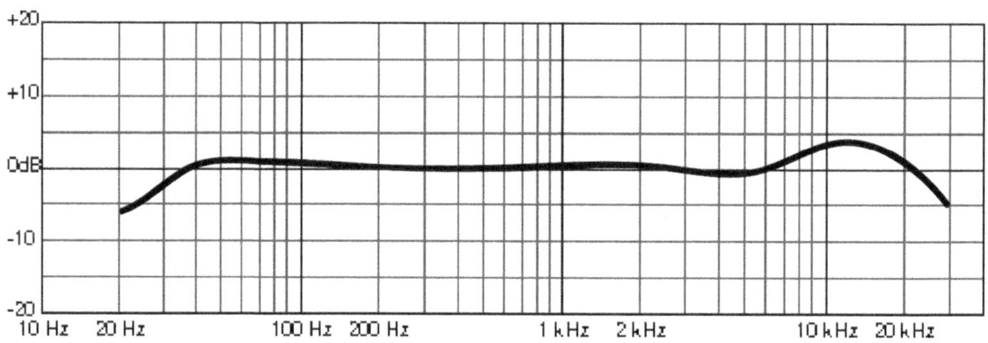

Shotgun microphone frequency response

The frequency response of a shotgun microphone rolls off dramatically below 100 Hz, which is below the range of speech, and has a 5 dB boost around 15 kHz. Because the shotgun microphone has been extensively used throughout the history of film sound, I consider this 15 kHz boost the clarity that is expected in this genre.

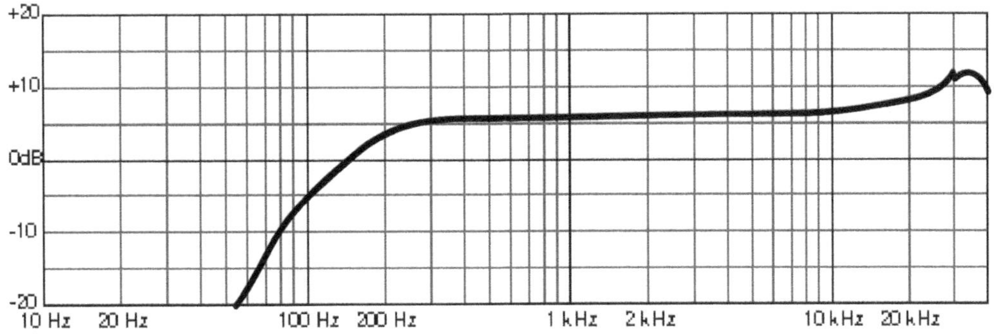

iPhone, iPad microphone frequency response

While not the best resource for recording audio for film, the microphones on devices look similar to a shotgun. In desperate, last-minute situations, I have had actors record pick-up dialog lines from across the country on their cell phones and email them to me.

AUDIO INTERFACE—AGAIN

Now that we've had a brief primer on how digital audio works, we can get an idea of what our audio interface will do.

It will first take a microphone's electric signal and convert it to line level using its preamplifier (preamp). It will then sample the converted signal using the onboard analog to digital (a/d) converter, which will change the analog signal into a sequence of zeroes and ones (binary code) that the computer will store. When you play back audio from the computer, the reverse will happen, only this time the convertor will be a digital to analog (d/a) convertor. You can see how important the components of the interface will affect the quality of the sound. Good convertors and preamps are extremely important.

Now you have to consider how many microphones you will need to connect to the computer at one time. Most audio post-production uses one microphone, which means that the interface will have only one main input. It is also possible that you will need up to four with the possibility of recording a few people at one time. A podcast would be a good example of this.

As you can see, the interface is as important as your computer and your audio software. When these three components are solid, you have an excellent DAW.

STUFF TO REMEMBER

- Mic/line level.
- Audio connectors.
- Microphone patterns, types, and frequency response.

PUTTING IT INTO PRACTICE

- Look behind the components you use for television, music, and gaming and identify the audio connectors.
- Record different types and patterns of microphones and try to identify each in a blind listening test.
- Research current audio interfaces and determine whether their features are appropriate for music production or audio post-production.

4 ORGANIZING OPEN MEDIA FRAMEWORKS (OMFS) (OR ADVANCED AUTHORING FORMATS (AAFS)) Planning the Tracks: Where to Start, and Organizing Your Work

When we receive audio media from an editor, we get to see how they think, their organizational priorities, and how important they consider the soundtrack to be in the final product. Are they organized individuals, or do they keep a "messy room"? Some of the best organized audio that I have worked with came from an editor who started her career editing sound on magnetic tape using razor blades. Some of the worst organized media came from an editor who told me it was my job to take care of the sound. Most editors are conscientious and ask me if they should organize the tracks before they send them to me. In these instances, I tell them to just leave it the way it is since I will organize the tracks anyway. Whatever the situation, we now have the opportunity to fully see what is going on. It's time to get organized.

Now that you have a comfortable place to work, and more importantly, a place to hear, it is time to start on something real. Typically we receive an OMF or an AAF from the editor. These are files made from the editing program that are used to transfer timeline information from one program to another. AAF is a newer format, but either works. For our purposes, we are interested in the audio portion (the bottom) of the editor's timeline. The picture comes as a separate file.

OMFs and AAFs can have the audio embedded in the document or in a separate folder. What is important is that we have a continuing dialog with the editor so that they know what we require. It is also wise to do a test delivery of audio; maybe just a small scene could be sent so that we can see that everything opens correctly. When software is updated, we can usually predict that there are bumps in the road whenever a new version is released, so it is always good to test the workflow before deadlines approach.

We assume that when we start the mixing process, the picture is locked. This means that the editing of the image is completed and no further changes will happen. This is rarely the case, however, for good reasons and bad. A good reason could be that the director or producer had a small screening and decided to make some adjustments. Improving the film is always justified. More often than not, though, I have found that the following scenario happens. Making a film is a monumental collaborative task that will consume many hours and dollars. Through great effort, the filmmaker has finally gotten their project to the point where he can pass it off to audio post, color correcting, and a variety of other finishing disciplines. Now, after being consumed by the project, the filmmaker has idle time on his hands. But he has to do something to make the film better. And now he starts playing with edits; a frame here, a frame there. I understand that it is difficult for him to let it go, but every decision he makes after handing the project over to the finishing departments will have significant consequences in time and budget.

OPENING OMFS OR AAFS

When the OMF or AAF is opened in our audio program, we sometimes have an option to link or to copy to the audio that is part of the OMF or AAF. If we link the media, our DAW will always refer back to the OMF or AAF, and consequently they must always stay with the project. Copying the audio rewrites all of the files into our program. I suggest that we copy the audio.

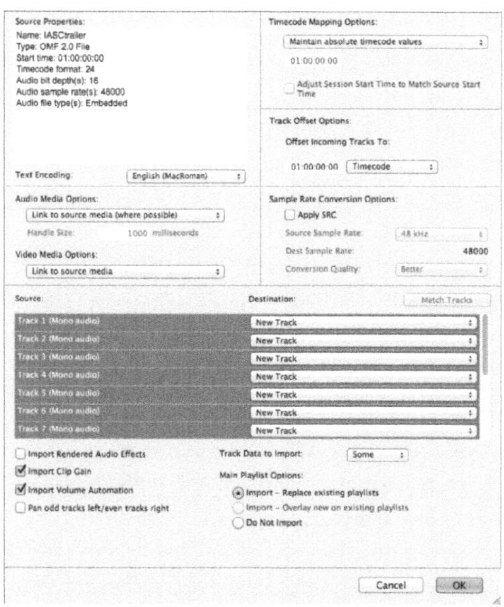

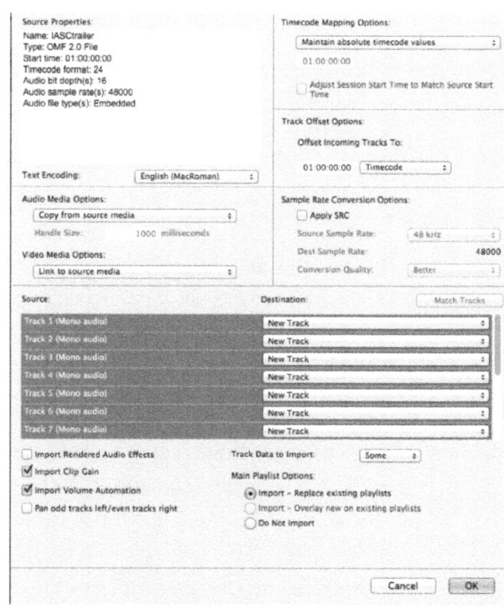

Why? First, copying the OMF instead of linking to it will generate all the audio necessary into your project folder. Second, some DAWs will not allow handles to open when linking.

Handles

Let's talk briefly about handles. Handles are the extensions of audio files that cannot be immediately seen. They can be dragged open or trimmed, and are very useful in crossfade.

In this example, we can see the cut on the top audio track.

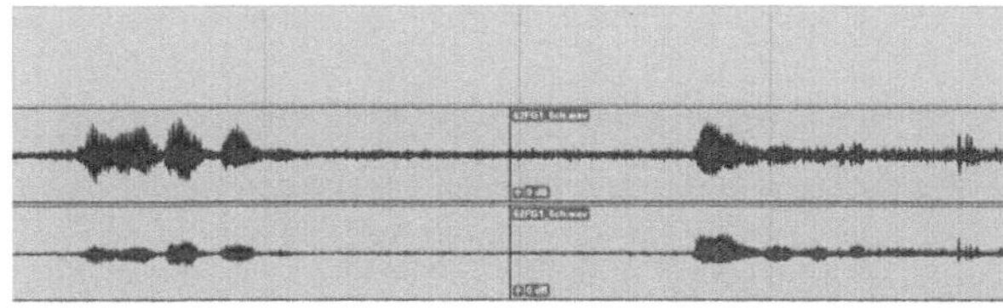

If we drag the audio on the right to a new track,

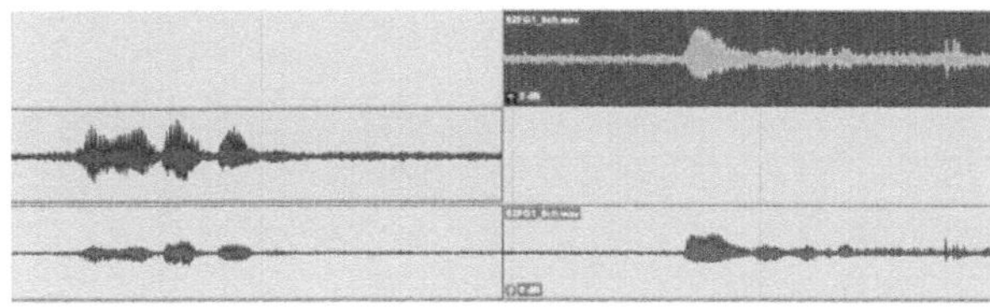

we can extend the audio past the edit point,

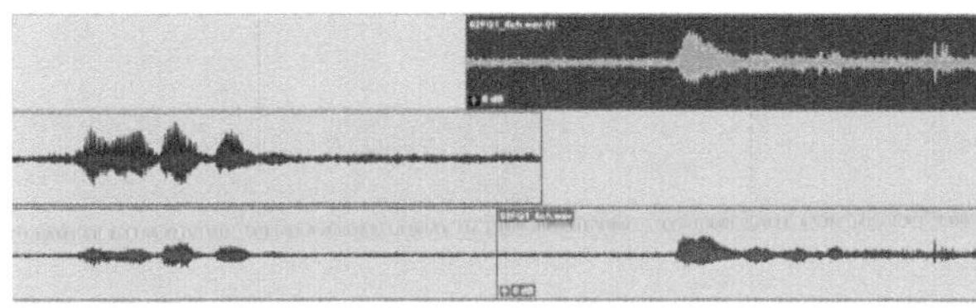

and add a fade on each individual file.

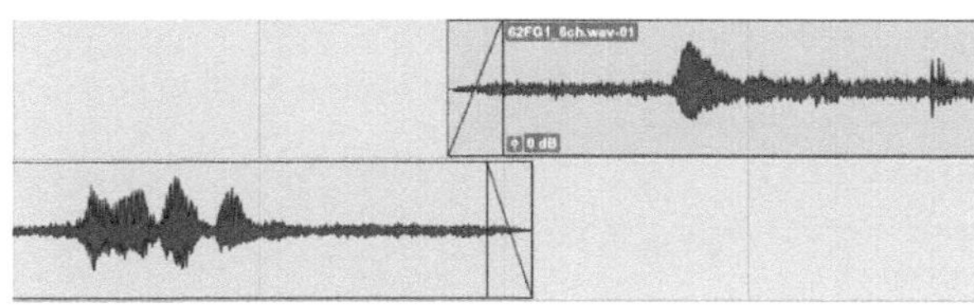

If the handles were non-existent, there would not be any audio to perform the fade.

Generally the editor makes the decision about how long the handles are to be from their editing program, when making the OMF or AAF. Usually the default time is one second on either end of the audio file. Sometimes I have received ten-second handles, but I have rarely needed more time than a second.

Let's open up an OMF and see what it looks like. Here is what the trailer of a documentary called *I Am Santa Claus* in Pro Tools, looking at a waveform view in 4.7, and the volume automation the editor used in 4.8.

And here's the same trailer in Adobe Audition (4.9). They look the same.

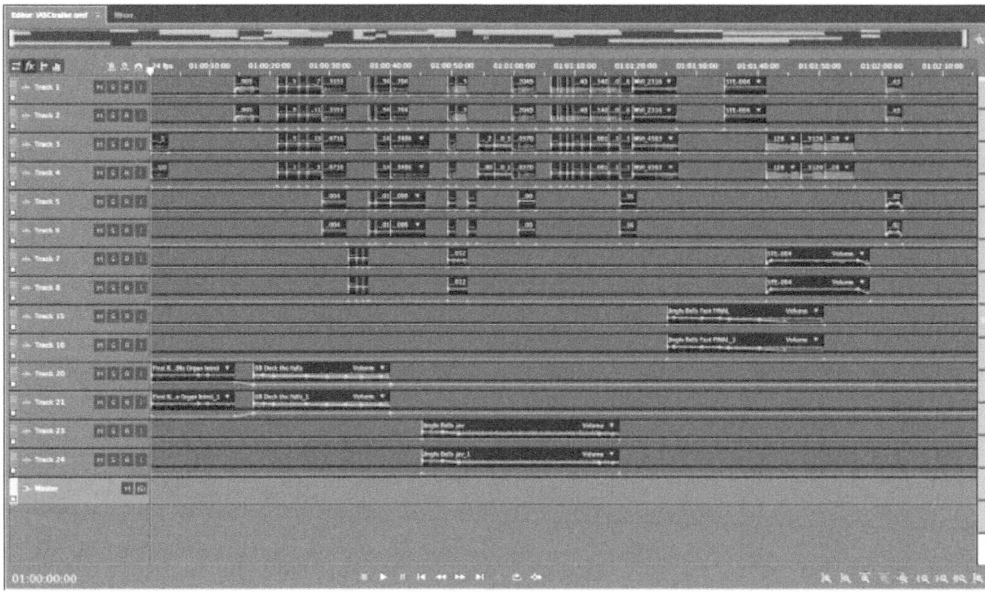

A Video Copy

Since we are editing and mixing some sort of media, whether it is a movie, video, webcast, or anything with an image, we are going to import the video into our session. Other than the obvious reason of having an image to react to, having the audio from the picture available at all times is essential for a couple of reasons. First, we have a reference of how the editorial department, and quite possibly the director and/or producer, heard the film. This can give us an idea of how they perceived relationships between dialog, music, and effects. Maybe it is a scene where just music is driving the scene and there is no ambient sound or dialog. There are many possibilities like this. Having the audio from the edit lets us hear what they were going for. A second, and more technical, reason is that we can always check that the audio from the OMF is in sync with the picture. If there are any errors from the OMF, we have a point of reference. Or, if we have slipped a piece of audio accidentally, we can place it where it belongs using the reference audio.

Often, the start of the picture changes. For example, editorial may have decided they will need a couple of seconds for a distributor's logo and we see that our audio from the OMF is early. We can slide all of the audio to the left and place it where it belongs. Also, when a "locked" picture is changed, we can use the new re-edited video and its audio to update our timeline.

Save a Copy

Now that everything is in front of us, we are ready to get started. If our project is called "The Great Film Mix," we should save it as "The Great Film mix yourname." Why? Before we start moving, rendering, and deleting any audio, we have an untouched version of the project to go back to when we make an error. And we will make an error—maybe we deleted a file and didn't realize it until weeks later—but now we will always have an untouched version to go back to.

Finally it is time to do some real organizing. Our goal is to get our session to the stage where all the audio can be easily identified. We also want to get the number of tracks down to the smallest size possible while keeping enough flexibility to make any change to any audio while mixing.

There is no right or wrong to this process and every project is unique. Here are some specific guidelines in achieving our organizational goal.

Remove Blank Files

The first order of business is to remove any unnecessary audio files. Often we will see audio files that *appear* to have no waveform, and they usually do not. They can be deleted.

Here is a scene from *Stray,* a feature film that I scored and mixed. You can see the blank files that show no waveform.

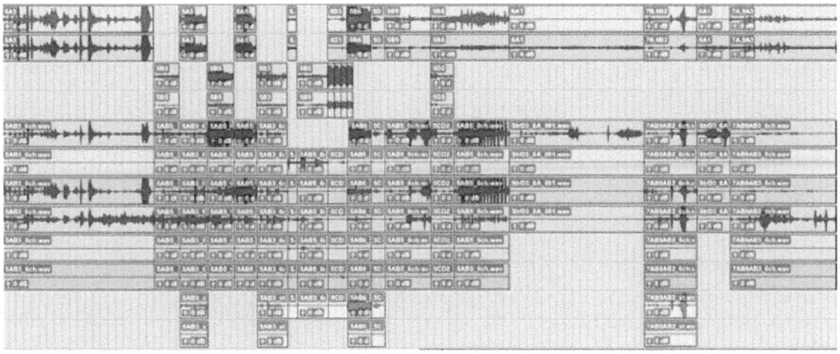

And with the blank files removed, it looks like this.

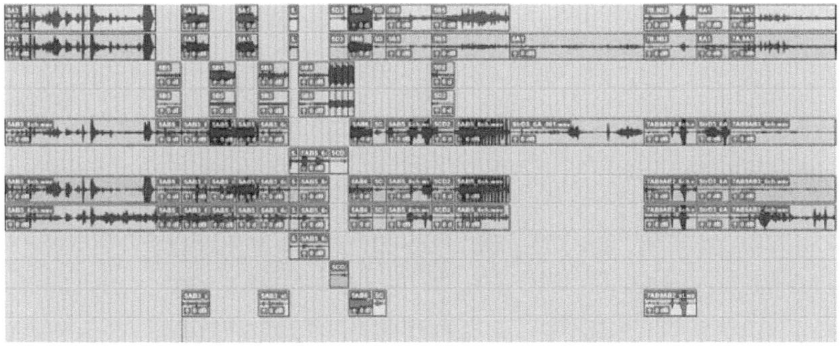

Then there are waveforms that are so low they cannot be used unless their levels are significantly increased. My opinion is that if the audio recorded is this low, it is probably because of a bad connection, an unused receiver, or a bad cable. Either way, it is unusable in our mix. Delete it.

Remove Dual Monos

Because the way audio is sometimes imported into a video editing system, in certain cases we get a doubling of audio track. It may well be, for example, that the video editor has imported a mono audio file into a stereo audio track, meaning that the program must double the audio to fit the slot. This generates unnecessary redundant audio tracks, which should be deleted, but how can we be sure they are identical copies and not a similar-looking audio file? We do not want to get rid of something that may be useful later.

In this example from the trailer of *Wish You Were Here*, we have two tracks that look identical. I question that these are identical tracks since both are labeled "VO Aaron." (Most voice-overs are recorded with one microphone and so it is unlikely the tracks are different.)

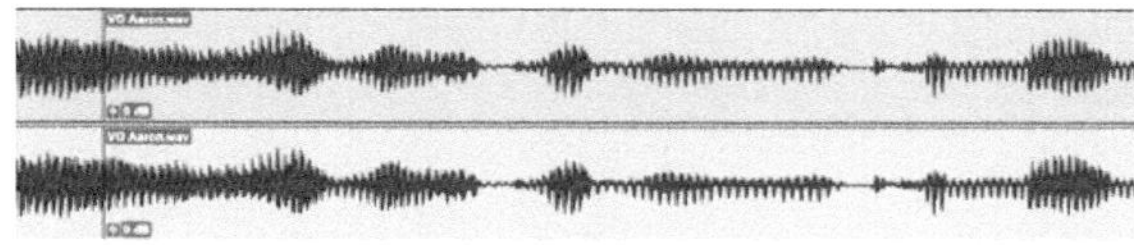

If we zoom in, they certainly *look* the same, but how can we be sure?

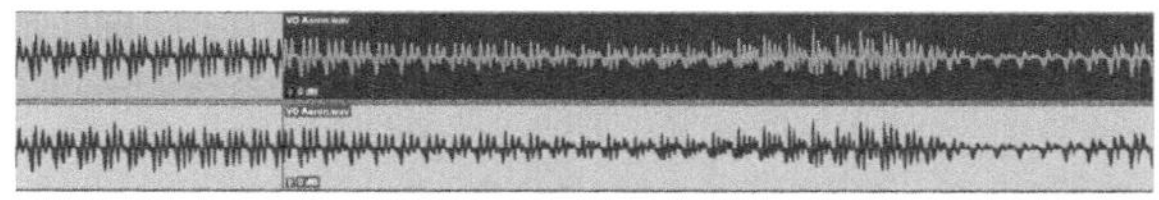

One way to be certain is to take one of the audio files and invert it. On this occasion I rendered (rewrote) it as a new file. You can see that one file is now a mirror image of the other.

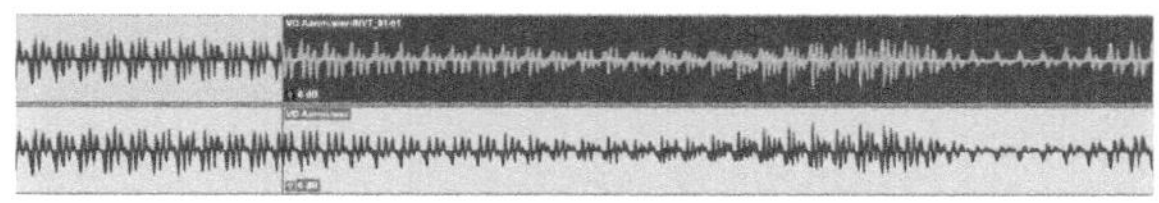

Here is a closer look.

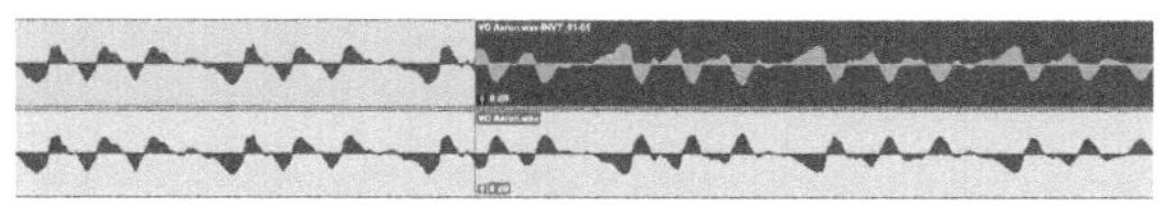

Now, if we play back the two files to listen to them together, *and* both are panned to the middle, we will hear nothing. The files have cancelled out each other, meaning they are identical and one of them can be deleted.

Sometimes we can use a plugin that can invert the file at a mouse click. In Pro Tools, it is in the Channel Strip plugin and the button looks like the Greek letter phi.

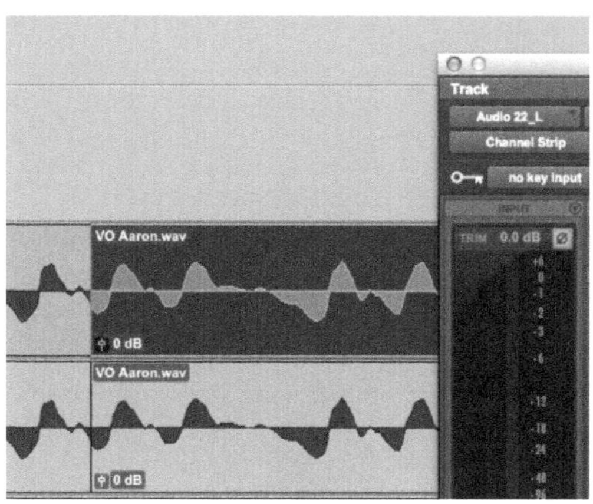

Φ

Greek letter phi

Define the Source of Audio

I then place audio onto tracks based on the microphones used. In the same scene from *Stray* (above), I know from the editor that the first four tracks (in red) are camera mics, so I can keep them where they are. Fortunately the editor was good at organizing, and I have an open dialog with him and can ask him any questions concerning his organization. I probably will not use many of the tracks from the camera mics, but for now, do not delete them.

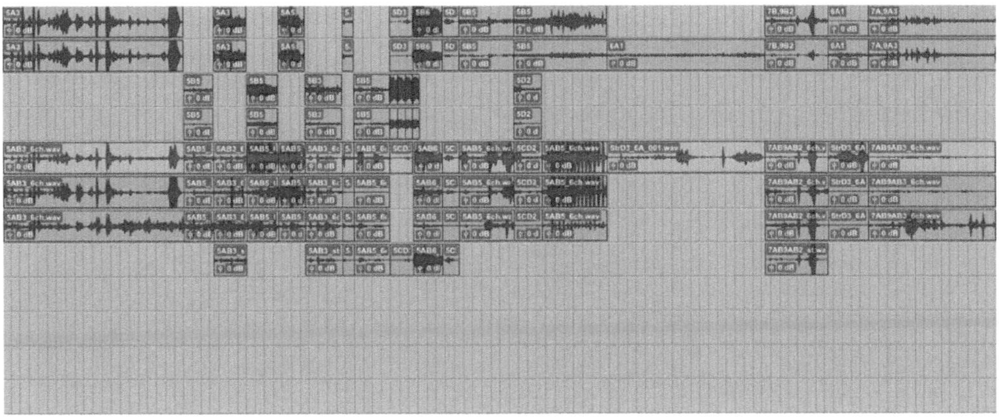

I then group the remaining audio based on whether it is recorded with a boompole mic, or whether a lavalier wireless was used. These decisions are made based on the way the audio sounds. Lavalier mics tend to be "bassy," and have little or no room reverb. Microphones used on a boompole will have more presence and have a room reverb that is consistent with the camera perspective. In this example, the boompole is in blue, and the lavaliers are in green.

Comparing the original OMF in figure 3.10, we can understand all of the audio in the scene just by glancing at the timeline, and I've reduced my track count by 4 while not losing any usable audio.

NDP

Next, go through a pass of the entire session. We are looking for audio that has no dialog, such as someone walking into a room, or an exterior street scene, or even somebody sleeping and snoring. (I do not consider snoring dialog and would include the audio in this group.) Move the selected audio to its own exclusive track. Often you will need more than one track, but two usually is enough. Label these tracks "NDP;" actually you can call them what you want, since I coined the term "non-dialog production."

CONSOLIDATE STEREO NON-DIALOG TRACKS

We then start looking for music and/or sound effects tracks that are stereo, but live on two mono tracks. This will help us to identify the tracks easily and to eliminate the risk of moving one of the tracks and not the other, which would put them out of sync.

In the case of the same scene in *Stray*, the yellow tracks are temporary music (often referred to as "temp" music) placed in the scene. The two mono audio files are moved to a stereo track, and we again save ourselves another two tracks.

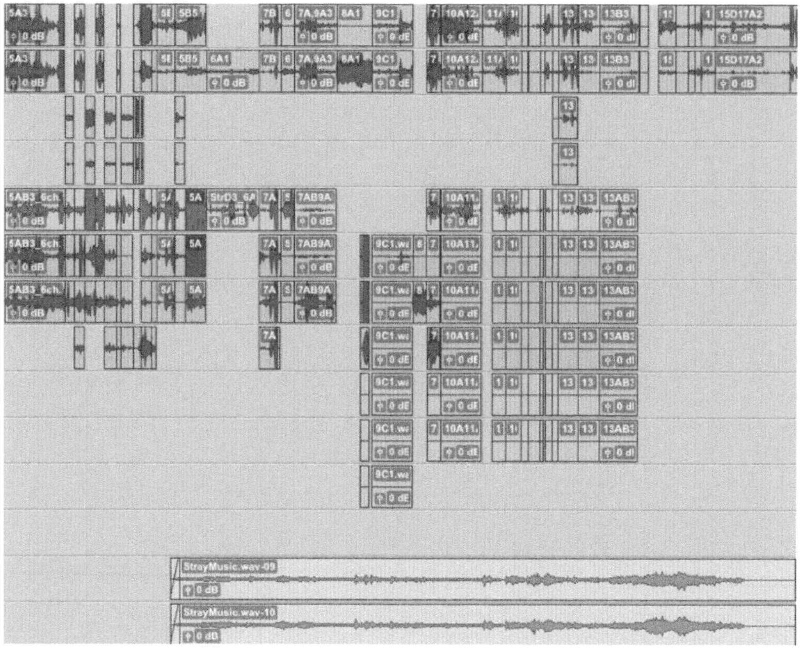
StrayMusic.wav-09
StrayMusic.wav-10

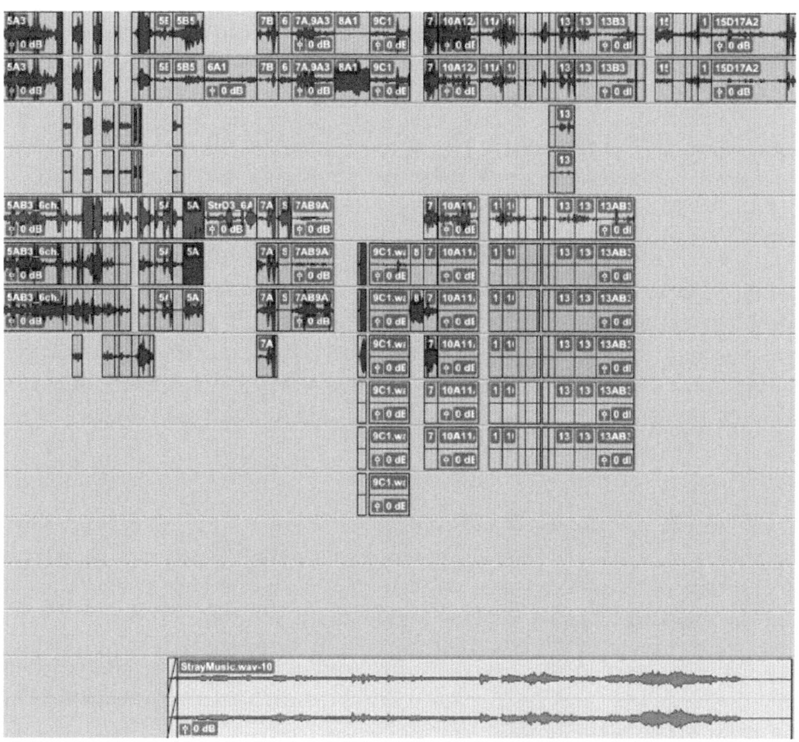
StrayMusic.wav-10

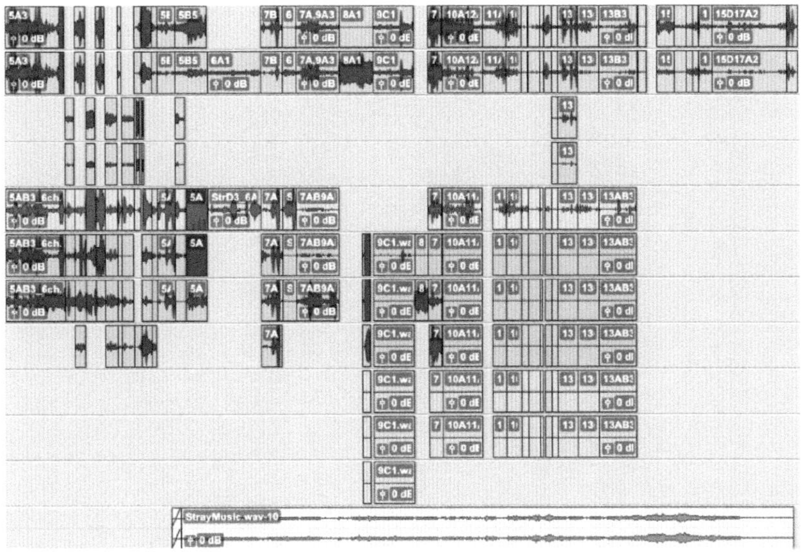

I cannot stress enough the importance of organizing audio sessions in such a way that the entire session can be understood at a glance. We may be working on many projects and revisiting some of them after a couple of weeks. Clearly organizing and labeling our tracks is the best way to efficiently use our time. This amazing technology now available to us seems to have compressed production schedules, so staying organized is the best way of staying sane.

STUFF TO REMEMBER

- Keep an open dialog with the editor.
- Save a copy of the audio mix session.
- Remove unnecessary audio.
- Organize audio tracks into Dialog, Effect, and Music sections.

PUTTING IT INTO PRACTICE

- Get familiar with inverting any type of audio on your DAW.
- Try inverting one track (left or right) of a stereo music track. When both play back in mono, what do you hear?
- If you are not working on a project, ask an editor to give you an OMF/AAF from a past project that he or she has worked on. Request an embedded/encapsulated one and another where the audio is in a separate folder. Try opening and organizing.

---

5 NARRATION

WHAT IS THAT VOICE AND WHERE IS IT COMING FROM?

Narration, also known as voice-over, is a disembodied voice that informs the audience what is going on in the story and moves it forward. As a form of storytelling, narration is the quickest and most efficient way to get details to the audience. Most narration is recorded after the picture is edited and is recorded to picture and not part of production sound. For this reason, I consider narration recording as part of audio post-production.

A filmmaker that I worked with for many years once told me that if narration is needed in a film, it is a sign that there is a problem in the storytelling, and the narration is the cheapest, or laziest, way of fixing the problem. I am still not sure whether I believe this or not, but the concept has stuck with me for many years, so I think that there is something to the statement. Yet narration can be extremely effective in films, as *Goodfellas* demonstrated. In this movie, the main character, Henry Hill, is describing his life in the mafia. While there is plenty of diegetic sound in the film, the movie is driven by narration. When we get to the end of the film, he breaks frame and looks into the camera, revealing that he has been telling us this story the whole time.

Another well-known example is *A Christmas Story*, where the writer, Jean Shepherd, drives the tale of his biographical anecdotes in a believable way that no on-camera actor could replicate.

Television shows also use voice-over well, as we can see in shows such as *Arrested Development*, *Scrubs*, and *Desperate Housewives*.

PLANNING A NARRATION RECORDING SESSION

Let us now plan our narration recording session. What type of narration is it? It could be for a documentary, a narrative film, or ever just a few wild lines—dialog recorded without camera to be inserted later in post production—that help move the story along.

In many cases, temporary narration is recorded by the editorial department, usually using a microphone plugged into the editing computer, recorded in the room with the editor. This is less than ideal—usually the computer is in the room and the fan noise will be difficult to remove—but the director or producer knows that this is a temporary solution and the narration is used as a placeholder to establish timing and pacing of the picture.

Most narrations are recorded in a professional studio with the voice actor (the "talent") working inside a vocal booth or other very quiet environment. The voice has to sound pristine, without any sense of space. For the last ten years, I have recorded narration in the same room as my mixing and composing setup. Usually the voice-over talent is just a few feet away from me, yet the recordings are successful because of where I place the microphone and how I treat the acoustics of the room. However your space sounds, there are some simple solutions that will achieve successful results.

- Place the talent and microphone stand on a rug. While minimizing any reverberation in the room, the rug will greatly reduce any foot noise created when the talent is moving or shifting their weight. That said, if your voice actor is standing on a rug on a hard floor surface such as wood or tile, you might consider placing some blankets on the floor to minimize any reverb in the room.
- The surface behind the talent should have some sound absorbing properties. Since the microphone typically used for this work has a cardiod pickup pattern, and thus picks up in the direction of the voice and not the entire room, it will also detect any sound behind the talent. In my space, I have sliding glass doors behind the talent, which is the worst possible surface for recording voice-overs, unfortunately. Knowing this, I have installed thick curtains that I pull over to cover the windows during recording. These prevent any sound from bouncing off of them and into the microphone. And since they are not a flat surface, sound is also diffused, or scattered.
- The talent will need a music stand to hold the script. Mine is a sturdy metal version, but it does vibrate, and it is possible to hear that vibration. It also becomes another surface off which sound will bounce. To solve the problem, I have cut a piece of thick carpet the size of the stand, and placed it so that the script can lie on top of it. In watching "behind the scenes" videos of professional voice-over sessions, you may notice that they use music stands that are made of vertical slats of wood—no bouncing sound, and no vibration. Some day I have to build one of those.

MICROPHONE CONSIDERATIONS

Generally, narration is recorded with large diaphragm cardiod condenser microphones. Condenser microphones reproduce a relatively flat frequency response, and the large diaphragm microphones deliver a more natural, accurate bass response than a small diaphragm mic. I am generalizing here. There are some excellent dynamic microphones that do a wonderful job at recording narration. Dynamic microphone's sensitivity is not as responsive after two or three feet away from the diaphragm, but if your space has a bit too much reverberation, this may be a better choice for you.

Today, you can find first-rate microphones at quite reasonable prices. If possible, try and borrow, or rent a variety of microphones to try out before you purchase one.

Placement

Now you have your recording setup ready to go, it is once again time to listen to your room. Place your microphone where you think you want to record, and record the room tone without any voice. Then, without changing the recording levels, move the microphone to another spot and record some more room tone. And then try other places in your space, even turning the axis of the mic 90 degrees. Listen for the quietest placement. Going through this process will teach you about your room. Our ears perceive sound differently than microphones. And most importantly, do this well in advance of the recording session. Once your talent arrives, everything should be in place and ready to go.

CASTING

The voice over/narration recording market has changed in recent years. If voice artists intend to stay in business long-term, they usually have to have their own recording setup at home. Many casting services are online and potential clients choose from a wide selection of demo reels and then direct the session over the phone. The talent then uploads the files for online delivery.

Often I am asked to recommend someone for a narration session, usually for a long industrial training video. Time is money. And since the time that you spend recording and editing the read directly affects the cost, choosing someone with experience makes the experience good for all involved. There are usually two ways of billing for these sessions. One is by the hour, where the client pays the talent a fee for their services, and you, the studio, buy the hour for the recording and editing of the read. In this case, the more efficient your talent is, the happier your client is. The other way of billing is a flat rate, where the client pays you one price that includes the fee of the talent *and* your fee for recording and editing. If you plan well and have competent talent, everyone is happy. If you do not plan well and the talent struggles in delivering what the client is looking for, however, or makes frequent errors in reading, you have to spend more time recording. Plus, you will have to put in the hours of editing the recording to deliver an acceptable product to your client. Agreeing to a flat fee can be risky, but if you know what you are doing and you are proficient at your craft, your expectations of time will be realistic.

Here is an example of a spot (commercial) where I had to cast a voice talent. The client wanted the narrator to sound like the Ronald Reagan reelection campaign spot from 1984 called “It's morning again in America.” Try a little experiment with yourself. Go to any voice-casting website and listen to the sample reels of voice over talent, and see if you can pick a talent that has the same feel as the Reagan spot.

https://www.youtube.com/watch?v=EU-IBF8nwSY

SCRIPTS

Most of the time, clients provide the script, and often there are last-minute changes. If they are bringing printed copies, the minimum numbers that I request is three; one for me, one for the talent, and one for the client. If possible, plan ahead and ask for an electronic version to be emailed to you in advance too. This allows you to not only print your own copies, but also to sort out any font size or spacing issues that your client may not have considered.

For longer reads, I like double spacing with a minimum 10-point font size. Make sure that page breaks happen at the end of the paragraph. Since you are going to pick up the page turn in the recording, having a page turn at the end of a paragraph will allow plenty of time for the talent to pause at the end of a thought, and then turn the page.

For shorter reads, such as a commercial, I leave plenty of space between sentences and even phrases that are fragmented sentence. Phrases in commercials are usually standalone thoughts and do not have to follow grammatical precedent. I also number the lines in the margin of the page. Even with leaving plenty of space and number margining, I am able to get the spot onto a single page.

Here is an example of a commercial script.

01 *Your IT is critical for your organizations to fulfill their missions.*

02 *You are constantly adding new functionality and leveraging new opportunities, in ever-changing worlds of day-to-day operations, applications, and innovations.*

03 *Your IT is on an evolving journey through an accelerating and demanding technology universe.*

04 *Wherever you want to go,* ***The Greatest Company Ever*** *is with you for that journey.*

05 *Three constellations of services provide you with the expertise to evolve, implement, and manage your* ***Greatest Product*** *environments, making your voyage to future easy.*

06 *Whether you need specific* ***Greatest Product ABC*** *or* ***XYZ*** *skills, or want to outsource all your IT headaches in order to concentrate on growing your business – or anything in between –*

07 ***The Greatest Company Ever*** *is with you.*

08 *We can help you evolve and further enhance your current applications, to capitalize on new market dynamics, and keep your* ***Greatest Product*** *environment up-to-date with the latest innovations.*

09 *We are from the center of the* ***Greatest Product*** *universe, we are the people who created* ***Greatest Product*** *technologies – we are THE experts. We are* ***The Greatest Company Ever.***

10 *We can innovate and build the future with you, while maintaining the high availability, security, stability, and business continuity you need and expect from* ***The Corporation.***

11 *This is your voyage to the new worlds of your digital enterprise, and wherever you want to go,* ***The Corporation*** *is with you for the journey.*

12 *Because, with* ***The Greatest Company Ever,*** *the sky is no longer the limit.*

WORKING WITH TALENT

Whether the voice talent is a professional or someone who is recording for the first time, your role as engineer is to make them as comfortable as possible. Generally, narration is recorded standing up. However, if the read is rather long, make sure a chair is available in case they wish to sit. And always have water available.

Now you are all ready to go and all technical aspects of the recording have been solved, prepped, and prepared prior to the talent, and usually the client, arriving at your studio. Once everybody has been introduced each other or caught up on current news and jokes, it is your responsibility to move the session forward. And it is important to do this without being pushy. I find myself always holding back. I want to finish and get everyone out of my place. Most of the time I get paid a flat rate, meaning that the faster I finish, the more I am compensated. I am not selling time, after all. I cannot let everyone know this, though, so I have developed certain people skills of making sure everyone is comfortable while moving everything along at a decent pace.

The recording has started. You are as responsible as the client in ensuring that the script is recorded correctly. And, while the read is recording, errors will inevitably crop up and need to be corrected. You first impulse will be to stop the talent and to make them aware of their mistakes, but I suggest waiting until the end of the sentence or paragraph instead of jumping on the mistake and interrupting. Most of the time, seasoned talent know that they have made a mistake, but they want to get through the thought. (After all, it is probably the first time that they have seen the script.)

And as with actors, do not read the script to them. If your client is not present, direct the talent, if necessary, by telling them to speed up or slow down the pace, or lift it up, or bring it down in feeling. The only time I *have* read lines to voice talent is when they are children. They may not understand the subtleties of what is necessary, but are excellent at mimicking what they hear.

I did a small web commercial about a child's product called "Monster Spray," where the client wrote the script and a brief melody at the end that needed to be read and sung by a little girl. Naturally, I enlisted my six-year-old daughter; she was free, and had her repeat exactly what I said and how I said it.

MARKING SCRIPTS DURING RECORDING

While the recording is happening, and you are listening to the read to make sure that every word is correctly delivered, you should be also planning on a way to edit it efficiently. The best way to do this is by leaving marks on the script to indicate issues such as:

- any errors, or where a re-read or restart takes place;
- which takes are the best ones;
- a page break where an extended pause needs to be closed.

Here is an example of a marked script and the reasons why I marked them.

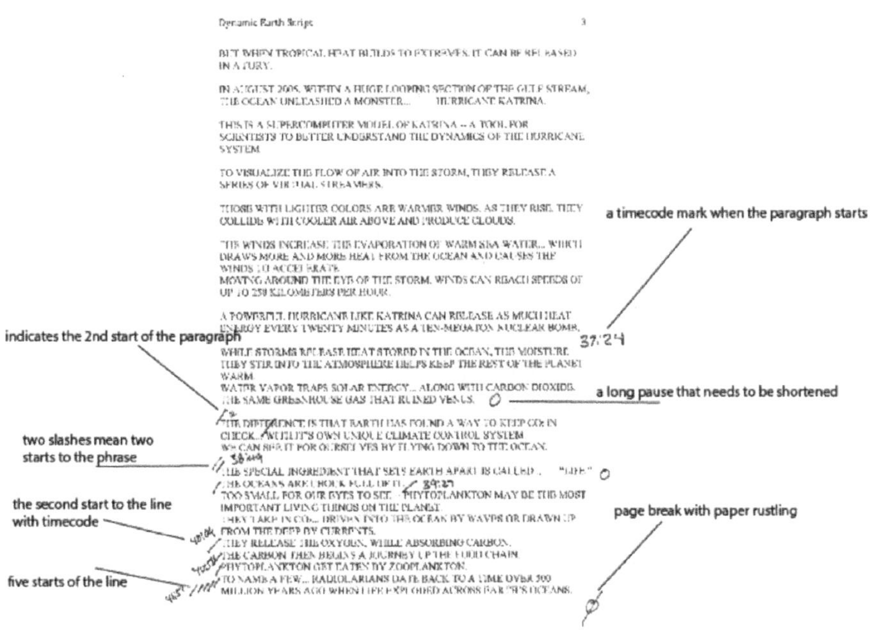

EDITING NARRATION

Now that the recording is finished and the clients and talent have left, it's time to edit the read. Clients rarely stay around to watch you edit, and it is usually uploaded to a website by the time they arrive at their next destination. It's a good idea to edit as soon as possible while the session is fresh in your memory.

For a narration like the above example, which was for a planetarium presentation, the running length of the read is not as important as it might be were it created for broadcast. We do not have to be concerned with the running time so we can make edits in a "natural sound" way, by leaving the spacing of phrases and breaths as they were recorded. In this edit, I am removing a bad take, yet I am keeping a little breath in front of the sentence. This will leave a natural space between the previous sentence and the connected correct sentence.

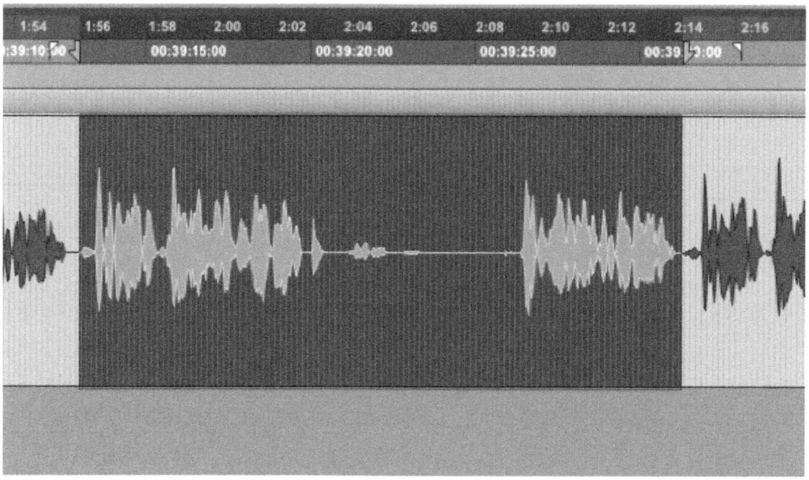

Inevitably, mistakes are made and discovered only after the talent has left. In this example the celebrity talent misread "nuclear" and read it as "nucular." And, after the George W. Bush issue (https://www.youtube.com/watch?v=hORaebYWDwk), it is something that had to be corrected.

In this edit, the word "clear" was edited into the "cu" on the erroneous read.

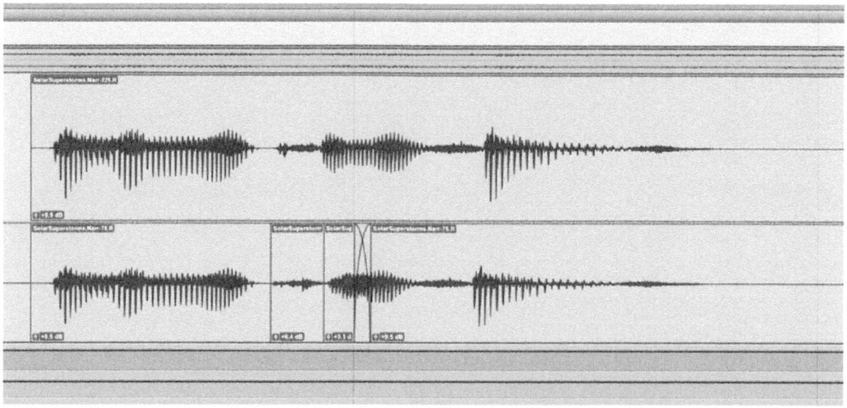

TIMED NARRATION

Often temporary narration is recorded during the editorial process. This is usually not recorded well—and often read too fast—but the real narration is recorded once the picture is locked. As a result, the new narration must be of a specific length and this should be stated in the script. An example of a timed script is below.

A (01:01 -)

We are 250 miles above the Earth, about to board the International Space Station.

B (01:09 -)

To maintain this altitude, we must orbit around our planet at over five miles per second

C (01:19 -)

Life is very different aboard the station. Up here we experience a sunset every ninety minutes,

D (01:30 -)

followed by a setting Moon.

This script gives me the start time of the phrase and lets me know that the read has to be finished before the next phrase starts, so that there can be a natural enough pause in between the phrases. Often using the temporary rough narration is enough information to let me know the amount of time required. If the new read is a little too long or too short, most audio software programs have a tool called TCE (time compression and expansion), which either extends or shortens the time of the read without changing the pitch. The process *will* alter the sound of the audio if it is overdone, however, but with some experience, you will learn the limits of this valuable tool.

A CLEAN RECORDING

As the audio engineer for the narration you are recording, you are responsible for delivering a voice-over track that is free from any clicks and plosions. Plosions, or "P-pops," are very brief low frequency bursts that hit the diaphragm of the microphone. They typically occur with consonant letters like b, t, or p. Often they cannot be removed due to the distortion that was recorded, so it is good practice to avoid them by having the talent direct their speech just to the left or right of the microphone. Veteran voice-over talent are well aware of this potential problem and move their head just a little off axis when they delivery these consonant spikes. It should go without saying that any rustling of clothes or shuffling of feet must be removed before you deliver the final version of the narration; most clients want all breaths removed too.

STUFF TO REMEMBER

Running a successful narration recording session requires a certain amount of multi-tasking. The engineer is responsible for the following:

- creating a quiet, comfortable microphone setup for the talent;
- providing a microphone that records well in your specific recording environment;

- having enough scripts available for all parties present, printed in a way that is easily read with as few page turns as possible;
- focusing on the read to mark any errors that may happen in a way that is efficient for the editing process;
- delivering an accurately read script that is free from any extraneous noises.

PUTTING IT INTO PRACTICE

- To test your narration recording setup, record yourself reading demo scripts found online.
- Find a favorite commercial with a narrator, then try to find a similar sounding voice from an online casting service.

6 DIALOG EDITING Every Word Is Important

WHAT IS DIALOG EDITING?

Most people are unsure what dialog editing is, or what dialog editors actually do. The other crafts of audio post-production are clearly understood, like sound designing, or Foley, or the obvious music editing. But dialog editing? Surely it just means taking out the curse words?

In fact, the dialog editor is responsible for all of the sound recorded in production. It is their job to make sense of the hundreds, or even thousands, of individual audio files and present them so that the audience will continue to believe the magic of storytelling in film. Many picture mistakes can break the trust of a movie watcher. Bad acting, unbelievable makeup, or poorly executed visual effects can take us out the magic. In audio, poorly matched ADR, unrealistic ambiences, changes in dialog volume, or distorted and corrupted audio will pull us out of the experience. If everything is executed correctly, the dialog editor's job is unnoticed, and only the lack of success will attract attention.

Narrative films are driven by dialog, and if one word is not understood, the subsequent sentence may also be missed because the viewer is asking, "What did they say?"

GETTING STARTED

Chapter 4 discusses the basic organization of an OMF/AAF that was delivered from the editor. In this chapter we will go into more detail about selecting the correct phrases that clearly deliver every word to the audience. We will choose between washed out recordings and deliver audio that has punch and clarity.

But before you start, you must be certain that you are armed with the necessary ingredients to do your job. Other than the OMF file, you will need the following:

- all the production sound files. Depending on the film, this may be around 50 gigabytes, but of course the size is always different, depending on the film;

- a script with scene numbers;
- a schedule of the shoot, listing which scenes were shot on each day of production.

All of this information is invaluable to you when it comes to searching out alternate takes to replace unusable phrases, words, or even syllables.

You will also need the production sound reports. These will tell you what microphone is on which track and how many takes of a scene there are, as well as notes that the production sound mixer has entered. Here is an example.

FileID	FileName	Tracks	FrameRate	TimeCode	UserBits	Duration	SampleRat	Scene	Take	Note	Name1	Name2	Name3	Name4	Name5	Name6	Name7	Name8	Name9	Name10
20	4-05020.WAV	2	23.98	16:02:20:07	4051600	1.1m	48000	11	1		MIX	BOOMIE								
21	4-05021.WAV	2	23.98	16:07:21:11	4051600	1.2m	48000	11	2		MIX	BOOMIE								
22	4-05022.WAV	2	23.98	16:13:24:05	4051600	1.2m	48000	11	3		MIX	BOOMIE								
6	4-05006.WAV	4	23.98	11:53:25:11	4051600	0.6m	48000	12	1		MIX	BOOMIE	VICTOR	BOON						
7	4-05007.WAV	4	23.98	11:55:15:09	4051600	1.1m	48000	12	2		MIX	BOOMIE	VICTOR	BOON						
8	4-05008.WAV	4	23.98	11:57:15:17	4051600	0.9m	48000	12	3		MIX	BOOMIE	VICTOR	BOON						
1	4-05001.WAV	4	23.98	10:07:03:05	4051600	2.1m	48000	13	1		MIX	BOOMIE	VICTOR	BOON						
2	4-05002.WAV	4	23.98	10:14:39:15	4051600	1.8m	48000	13	2		MIX	BOOMIE	VICTOR	BOON						
3	4-05003.WAV	4	23.98	10:24:12:04	4051600	1.7m	48000	13	3		MIX	BOOMIE	VICTOR	BOON						
27	4-05027.WAV	2	23.98	17:22:51:17	4051600	1.3m	48000	16	1		MIX	BOOMIE								
28	4-05028.WAV	2	23.98	17:25:30:11	4051600	1.4m	48000	16	2		MIX	BOOMIE								
29	4-05029.WAV	2	23.98	17:32:37:18	4051600	1.4m	48000	16	3		MIX	BOOMIE								
30	4-05030.WAV	2	23.98	17:36:16:23	4051600	1.2m	48000	16	4		MIX	BOOMIE								
38	z001038.WAV	5	23.98	19:01:31:09	4041600	3.4m	48000	17	1		MIX	BOOMIE	NOVAK	BOON	FLETCHER					
39	z001039.WAV	5	23.98	19:12:13:05	4041600	2.9m	48000	17	2		MIX	BOOMIE	NOVAK	BOON	FLETCHER					
40	z001040.WAV	4	23.98	19:17:36:04	4041600	1.5m	48000	17	3		MIX	BOOMIE	NOVAK		FLETCHER					
41	z001041.WAV	4	23.98	19:20:03:22	4041600	2.3m	48000	17	4		MIX	BOOMIE	NOVAK		FLETCHER					
3	z002003.WAV	2	23.98	10:39:22:09	0	3.8m	48000	18	1		MIX	BOOMIE								
1	4-07001.WAV	7	23.98	11:20:49:02	4071600	3.0m	48000	20	1		MIX	BOOMIE	BOONE	REDMAINE	NOVACK	FLETCHER	SLATE			
2	4-07002.WAV	7	23.98	11:29:03:22	4071600	3.0m	48000	20	2		MIX	BOOMIE	BOONE	REDMAINE	NOVACK	FLETCHER	SLATE			
1	4-07001.WAV	7	23.98	11:20:49:02	4071600	3.0m	48000	20	1		MIX	BOOMIE	BOONE	REDMAINE	NOVACK	FLETCHER	SLATE			
2	4-07002.WAV	7	23.98	11:29:03:22	4071600	3.0m	48000	20	2		MIX	BOOMIE	BOONE	REDMAINE	NOVACK	FLETCHER	SLATE			
33	z001033.WAV	5	23.98	18:04:25:21	4041600	4.3m	48000	21	1		MIX	BOOMIE	NOVAK	BOON	FLETCHER					
34	z001034.WAV	5	23.98	18:14:17:02	4041600	2.2m	48000	21	2		MIX	BOOMIE	NOVAK	BOON	FLETCHER					
35	z001035.WAV	5	23.98	18:18:39:01	4041600	2.3m	48000	21	3		MIX	BOOMIE	NOVAK	BOON	FLETCHER					
1	z001001.WAV	5	23.98	11:06:44:20	4041600	1.1m	48000	24	1		MIX	BOOMIE	NOVAK	BOON	FLETCHER					
2	z001002.WAV	5	23.98	11:10:52:00	4041600	1.5m	48000	24	2		MIX	BOOMIE	NOVAK	BOON	FLETCHER					
3	z001003.WAV	5	23.98	11:12:22:05	4041600	1.1m	48000	24	3		MIX	BOOMIE	NOVAK	BOON	FLETCHER					
12	z001012.WAV	5	23.98	12:41:05:05	4041600	5.4m	48000	26	1		MIX	BOOMIE	NOVAK	BOON	FLETCHER					
14	z001014.WAV	4	23.98	13:11:31:05	4041600	1.6m	48000	31	1		MIX	BOOMIE	NOVAK	BOON						
15	z001015.WAV	4	23.98	14:29:18:15	4041600	5.8m	48000	34	1		MIX	BOOMIE	NOVAK	BOON						
30	z001030.WAV	4	23.98	16:37:20:11	4041600	1.9m	48000	37	1		MIX	BOOMIE	NOVAK	BOON						
31	z001031.WAV	4	23.98	16:40:03:09	4041600	1.4m	48000	37	2		MIX	BOOMIE	NOVAK	BOON						
32	z001032.WAV	4	23.98	16:42:49:00	4041600	1.3m	48000	37	3		MIX	BOOMIE	NOVAK	BOON						
22	4-21022.WAV	3	23.98	20:24:25:14	4211600	1.8m	48000	38	1		MIX	BOOMIE	CHALICE							
23	4-21023.WAV	3	23.98	20:27:25:09	4211600	2.1m	48000	38	2		MIX	BOOMIE	CHALICE							
1	4-21001.WAV	2	23.98	12:54:20:19	4211600	1.1m	48000	39	1		MIX	BOOMIE								
2	4-21002.WAV	2	23.98	12:58:27:04	4211600	0.7m	48000	39	2		MIX	BOOMIE								
19	4-13019.WAV	2	23.98	15:38:23:07	4131600	4.9m	48000	40	1		MIX	BOOMIE								
20	4-13020.WAV	2	23.98	15:46:27:15	4131600	5.1m	48000	40	2		MIX	BOOMIE								
25	4-13025.WAV	3	23.98	17:49:42:13	4131600	1.2m	48000	41	1		MIX	BOOMIE	PLANT							
26	4-13026.WAV	3	23.98	17:54:17:05	4131600	1.0m	48000	41	2		MIX	BOOMIE	PLANT							
27	4-13027.WAV	3	23.98	17:56:11:12	4131600	1.0m	48000	41	3		MIX	BOOMIE	PLANT							
1	4-18001.WAV	2	23.98	08:24:18:13	4181600	3.0m	48000	42	1		MIX	BOOMIE								
2	4-18002.WAV	2	23.98	08:27:20:03	4181600	0.6m	48000	42	2		MIX	BOOMIE								
3	4-18003.WAV	2	23.98	08:28:32:09	4181600	0.9m	48000	42	3		MIX	BOOMIE								
12	4-21012.WAV	3	23.98	15:03:01:15	4211600	2.8m	48000	44	1		MIX	BOOMIE	CHALICE							
13	4-21013.WAV	3	23.98	15:10:04:11	4211600	7.5m	48000	44	2		MIX	BOOMIE	CHALICE							
16	4-18016.WAV	2	23.98	11:03:51:18	4181600	1.4m	48000	46	1		MIX	BOOMIE								
17	4-18017.WAV	2	23.98	11:06:01:06	4181600	3.3m	48000	46	2		MIX	BOOMIE								
24	z001024.WAV	3	23.98	16:03:24:09	4041600	0.7m	48000	47	1		MIX	BOOMIE			FLETCHER					
25	z001025.WAV	3	23.98	16:04:31:19	4041600	0.7m	48000	47	2		MIX	BOOMIE			FLETCHER					
14	4-18014.WAV	2	23.98	10:46:02:17	4181600	1.0m	48000	48	1		MIX	BOOMIE								
15	4-18015.WAV	2	23.98	10:48:01:20	4181600	0.8m	48000	48	2		MIX	BOOMIE								
4	4-18004.WAV	2	23.98	08:33:59:09	4181600	0.9m	48000	49	1		MIX	BOOMIE								
5	4-18005.WAV	2	23.98	08:35:46:11	4181600	0.9m	48000	49	2		MIX	BOOMIE								
26	z001026.WAV	5	23.98	16:10:53:00	4041600	1.0m	48000	50	1		MIX	BOOMIE	NOVAK	BOON	FLETCHER					
27	z001027.WAV	5	23.98	16:12:36:08	4041600	0.6m	48000	50	2		MIX	BOOMIE	NOVAK	BOON	FLETCHER					
9	4-07009.WAV	5	23.98	15:49:50:22	4071600	2.6m	48000	51	1		MIX	BOOMIE	BOONE	REDMAINE	NOVACK					
17	4-07017.WAV	3	23.98	17:21:31:09	4071600	1.5m	48000	53	1		MIX	BOOMIE	BOONE							
18	4-07018.WAV	3	23.98	17:25:17:09	4071600	1.3m	48000	53	2		MIX	BOOMIE	BOONE							
16	z001015.WAV	4	23.98	14:38:03:18	4041600	2.5m	48000	54	1		MIX	BOOMIE	NOVAK	BOON						
16	4-21016.WAV	3	23.98	16:03:33:18	4211600	1.3m	48000	66	1		MIX	BOOMIE	CHALICE							
17	4-21017.WAV	3	23.98	16:05:43:00	4211600	0.6m	48000	66	2		MIX	BOOMIE	CHALICE							
18	4-21018.WAV	3	23.98	16:06:46:05	4211600	0.8m	48000	66	3		MIX	BOOMIE	CHALICE							
21	4-18021.WAV	3	23.98	12:56:10:16	4181600	1.5m	48000	69	1		MIX	BOOMIE					BOOM2			
22	4-18022.WAV	3	23.98	12:58:00:10	4181600	2.6m	48000	69	2		MIX	BOOMIE					BOOM2			
7	4-19007.WAV	4	23.98	12:51:51:20	4191600	1.1m	48000	70	1		MIX	BOOMIE	BRADSHAW	PLANT						
8	4-19008.WAV	4	23.98	12:53:59:15	4191600	1.3m	48000	70	2		MIX	BOOMIE	BRADSHAW	PLANT						
9	4-19009.WAV	4	23.98	12:56:00:19	4191600	1.4m	48000	70	3		MIX	BOOMIE	BRADSHAW	PLANT						
8	4-18008.WAV	2	23.98	09:51:36:08	4181600	1.2m	48000	74	1		MIX	BOOMIE								
1	4-19001.WAV	2	23.98	09:14:19:09	4191600	1.2m	48000	76	1		MIX	BOOMIE								
2	4-19002.WAV	2	23.98	09:17:30:12	4191600	2.1m	48000	76	2		MIX	BOOMIE								
18	4-18018.WAV	3	23.98	12:30:20:14	4181600	1.6m	48000	80	1		MIX	BOOMIE					BOOM2			
33	4-18033.WAV	3	23.98	16:26:29:18	4181600	2.0m	48000	81	1		MIX	BOOMIE	SALERNO							
10	4-19010.WAV	3	23.98	13:04:10:02	4191600	0.9m	48000	83	1		MIX	BOOMIE	BRADSHAW							
40	4-18040.WAV	3	23.98	17:13:19:18	4181600	2.3m	48000	88	1		MIX	BOOMIE		SEIG						
41	4-18041.WAV	3	23.98	17:17:32:16	4181600	0.9m	48000	88	2		MIX	BOOMIE		SEIG						
42	4-18042.WAV	3	23.98	17:19:12:00	4181600	0.9m	48000	88	3		MIX	BOOMIE		SEIG						
43	4-18043.WAV	3	23.98	17:20:13:09	4181600	1.2m	48000	88	4		MIX	BOOMIE		SEIG						
13	4-19013.WAV	3	23.98	13:24:26:05	4191600	1.4m	48000	89	1		MIX	BOOMIE	BRADSHAW							
24	4-18024.WAV	5	23.98	14:59:16:14	4181600	1.2m	48000	90	1		MIX	BOOMIE	SALERNO	GUARD			BOOM2			
25	4-18025.WAV	5	23.98	15:02:23:13	4181600	0.8m	48000	90	2		MIX	BOOMIE	SALERNO	GUARD			BOOM2			
26	4-18026.WAV	5	23.98	15:03:59:02	4181600	1.2m	48000	90	3		MIX	BOOMIE	SALERNO	GUARD			BOOM2			
14	4-19014.WAV	3	23.98	13:35:56:12	4191600	1.2m	48000	95	1		MIX	BOOMIE	BRADSHAW							
3	4-19003.WAV	3	23.98	10:16:58:23	4191600	7.5m	48000	97	1		MIX	BOOMIE					BOOM2			
15	4-19015.WAV	3	23.98	15:26:57:14	4191600	1.3m	48000	98	1		MIX	BOOMIE	BRADSHAW							
16	4-19016.WAV	3	23.98	15:31:42:14	4191600	1.1m	48000	98	2		MIX	BOOMIE	BRADSHAW							
17	4-19017.WAV	3	23.98	15:39:04:15	4191600	1.6m	48000	98	3		MIX	BOOMIE	BRADSHAW							
4	4-19004.WAV	2	23.98	11:22:05:16	4191600	5.7m	48000	99	1	NO IDEA	MIX	BOOMIE								
19	4-19019.WAV	3	23.98	16:17:48:15	4191600	3.1m	48000	103	1		MIX	BOOMIE	BRADSHAW							
21	4-19021.WAV	3	23.98	18:14:47:19	4191600	0.8m	48000	104	1		MIX	BOOMIE					BOOM2			
28	4-13028.WAV	2	23.98	18:32:08:07	4131600	1.1m	48000	106	1		MIX	BOOMIE								
9	4-11009.WAV	2	23.98	10:47:56:15	4121600	1.4m	48000	108	1		MIX	BOOMIE								
1	4-13001.WAV	2	23.98	10:06:11:06	4131600	1.3m	48000	109	1		MIX	BOOMIE								
2	4-13002.WAV	2	23.98	10:19:04:02	4131600	1.1m	48000	109	2		MIX	BOOMIE								
3	4-13003.WAV	2	23.98	10:24:08:21	4131600	2.1m	48000	109	3		MIX	BOOMIE								
6	z002006.WAV	6	23.98	15:22:28:14	0	1.6m	48000	111	1		MIX	BOOMIE	BOON	NOVACK	FLETCHER		BOOM2			
7	z002007.WAV	6	23.98	15:26:55:04	0	1.5m	48000	111	2		MIX	BOOMIE	BOON	NOVACK	FLETCHER		BOOM2			
8	z002008.WAV	6	23.98	15:30:35:12	0	1.3m	48000	111	3		MIX	BOOMIE	BOON	NOVACK	FLETCHER		BOOM2			
9	z002009.WAV	6	23.98	15:33:50:18	0	1.3m	48000	111	4		MIX	BOOMIE	BOON	NOVACK	FLETCHER		BOOM2			
10	z002010.WAV	6	23.98	15:36:47:03	0	1.5m	48000	111	5		MIX	BOOMIE	BOON	NOVACK	FLETCHER		BOOM2			
15	z002015.WAV	2	23.98	16:35:21:18	0	1.1m	48000	112	1		MIX	BOOMIE								
16	z002016.WAV	3	23.98	16:40:19:17	0	1.0m	48000	112	2		MIX	BOOMIE					BOOM2			
8	4-15008.WAV	3	23.98	12:34:09:21	4151600	1.2m	48000	119	1		MIX		SEIG				BOOM2			
9	4-15009.WAV	3	23.98	12:38:31:06	4151600	1.2m	48000	119	2		MIX		SEIG				BOOM2			
25	4-07025.WAV	3	23.98	19:27:07:00	4071600	1.0m	48000	120	1		MIX	BOOMIE			SEIG					
26	4-07026.WAV	4	23.98	19:30:46:15	4071600	0.9m	48000	120	2		MIX	BOOMIE			SEIG		BOOM2			
7	4-15007.WAV	3	23.98	12:08:12:05	4151600	0.7m	48000	124	1		MIX		SEIG				BOOM2			
25	4-11025.WAV	5	23.98	16:47:59:09	4121600	1.2m	48000	125	1		MIX	BOOMIE	BOONE	NOVACK	FLETCHER					
26	4-11026.WAV	5	23.98	16:49:35:20	4121600	1.1m	48000	125	2		MIX	BOOMIE	BOONE	NOVACK	FLETCHER					
27	4-11027.WAV	6	23.98	17:03:24:10	4121600	1.2m	48000	125	3		MIX	BOOMIE	BOONE	NOVACK	FLETCHER		BOOM2			
15	4-11015.WAV	4	23.98	14:31:04:01	4121600	1.6m	48000	126	1		MIX	BOOMIE				SEIG	BOOM2			
16	4-11016.WAV	4	23.98	14:34:45:17	4121600	1.3m	48000	126	2		MIX	BOOMIE				SEIG	BOOM2			
17	4-11017.WAV	4	23.98	14:39:52:05	4121600	1.4m	48000	126	3		MIX	BOOMIE				SEIG	BOOM2			
32	4-11032.WAV	5	23.98	18:01:24:21	4121600	1.4m	48000	127	1		MIX	BOOMIE	BOONE	NOVACK	FLETCHER					
1	4-11001.WAV	2	23.98	08:54:16:23	4121600	1.7m	48000	128	1		MIX	BOOMIE								
2	4-11002.WAV	2	23.98	08:58:10:10	4121600	0.7m	48000	128	2		MIX	BOOMIE								
3	4-11003.WAV	2	23.98	08:59:45:07	4121600	1.2m	48000	128	3		MIX	BOOMIE								
4	4-11004.WAV	2	23.98	09:02:05:16	4121600	1.1m	48000	128	4		MIX	BOOMIE								

115_T1_4-08001.WAV

From this production sound report, I find the line items that are labeled "WILD." Wild sounds are takes that are recorded without a camera, with the intent to record acceptable audio that was impossible to record when the picture was shot. There can be many reasons for this. It may be a line in a fight scene that cannot be heard, or a crowd cheering, or a softly spoken line that has to cut through a noisy scene. For a reason that I have yet to discover, the production sound team felt it was necessary to take the time in production to record wild takes. So, as I prepare to start editing, I have a list in front of me of solutions to problems that I have not yet found. I keep the list nearby and as I approach a scene in question, it may change the way I think about the scene.

TimeCode	Duration	Scene	Take
20:47:41:22	0.7m	108WILD	1
12:17:55:03	1.9m	109WILD	1
10:49:24:06	0.7m	112WILD	1
17:44:44:20	0.6m	11WILD	1
17:45:31:17	1.4m	11WILD	2
17:47:03:10	0.9m	11WILD	3
17:47:59:03	0.5m	11WILD	4
20:04:40:13	1.2m	120WILD	1
11:00:27:12	0.6m	126ALT-WILD	1
10:51:16:19	4.8m	126WILD	1
15:08:30:10	1.4m	126WILD	3
16:56:50:16	8.4m	40WILD	1
20:24:45:06	1.0m	70WILD	1
10:18:55:18	2.4m	74WILD	1
20:26:43:21	0.8m	97WILD	1
10:56:31:03	0.7m	WILD	1
10:57:22:12	1.9m	WILD	2

A list of available wild tracks for *Death House*

I get started in the process by doing what I call "the painful first pass." Depending on the project, I may go through this process when organizing the OMF, or I may do it after that organization. Either way, the task is the same, which is to make the audio sound as clear and consistent as possible. The reason I refer to it as "painful" is because the pass will reveal precisely what has to be done to finish the mix. During this stage I will decide which scenes will need Foley, what sound effects will need to be added, what phrases will have to be replaced, and a number of lists will be generated. This "painful first pass" will tell me the amount of work that has to be done, as beforehand all I can do is guess. On most of the independent films that I work on, I am the entire audio post-production department, responsible for dialog editing, ADR, sound design, composing music, and mixing the film in surround according to the deliverable specs of the distributor. The first pass helps me understand the scope of the work in front of me, and the level of shortcomings in production sound.

During the first pass, I work with headphones only. There is a lot of surgery that happens in splicing slivers of audio seamlessly together, and working with headphones is like a microscope for audio. Dialog editing is like solving problems and I want to hear all of them. The frequency response of headphones is full-range so I will be aware of hi frequency hiss or static issues. There is a tendency to put filters of some tracks, like a high-pass filter to eliminate low frequencies. The low frequencies are below what is heard as speech, so why not? Go without the filters for now, however, as we are not yet at a place, or maybe in an environment, where we should be making decisions about low frequencies. We have to *hear* the problems to solve the problems.

Even if music, sound effects, or ambiences have been added to the film and are in the timeline, I do not allow myself to hear them during this process: Any audio other than the dialog will only cover up problems that I am to resolve. I guess that's another reason why I call it "painful," because I'd like to hear those other elements to immerse myself in the story. But I won't do it. I find that if I can make myself happy with the dialog wearing headphones, the dialog will translate well into speakers. So there I sit, for days, making lists, searching files, editing minute slivers of backgrounds, so that the rest of the audio post process is creative and rewarding. If you are not the sole audio post team member, the work that you do here makes you invaluable to the storytelling process.

This is also the time to check sync. Your reference will be the audio that is on the supported video. Go through the film, scene by scene, to confirm that the OMF is in sync with the audio on the video. I always keep that audio on a track at the top of the session as a constant reference in case I accidentally slip something. Once you know that you audio is in sync you never have to worry about it again. If you are unsure of sync, it can niggle away at you for the rest of your edit.

ORGANIZING THE TRACKS

Every film is different and there is no one way to start organizing. I try to decide whether the boom mic will be the dominant source or the lavaliers (often called radio or wireless, or "lavs" for short). The boom mic is the preferred choice because it gives the scene a sense of space, and a good boom operator can actually mix the film by the placement of his microphone during the scene. If there is a lot of unwanted noise, or scene that have wide shots, lavs would be a better choice. Lavs can sound excellent yet hiding them under clothing can cause unwanted noise if fabric rubs on them.

When you start the edit, you probably will not understand the film's voice for a while. It will take some time to figure out the best way to organize the film. Usually the first few scenes may not be your best since you have not deciphered the sound of the film and it will have to be revisited. In fact some dialog editors start in the middle of the film because the beginning of the film is the most important in defining the sound of the film to the audience, and they choose to work on it later in the process when they have properly got to grips with the voice of the film.

You will notice that each audio region is numbered for scene and take. This is invaluable for tracking down alternate, as well as keeping track of where you are in the script. You will know when wild lines are coming up.

For each scene, I decide which mics I will use and mute out the others. I don't delete them from the timeline, though, in case I want to use them later. The muted audio tracks are a visual reference of what is available in the scene. By keeping them there, I know the option of changing my mind exists. If there is nothing but one track, like a boompole, I will know later on in the mixing process that there is no other choice. However, during this first pass, I want to make a decision.

SMOOTH TRANSITIONS

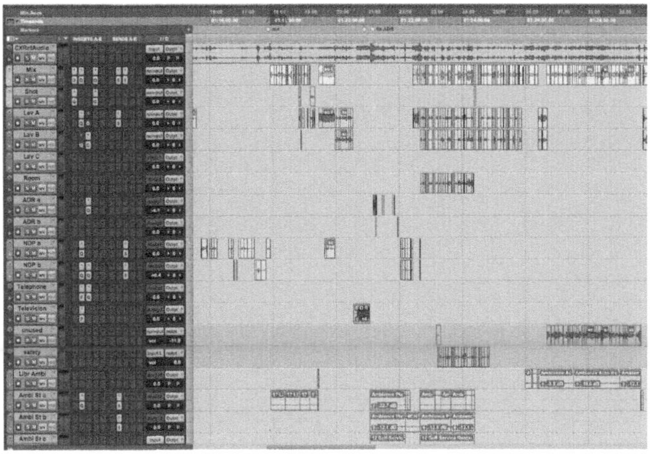

Making dialog decisions while editing *Death House*

Films are generally not shot in sequence. This means that as the scene cuts from one shot to another, a considerable amount of time has passed. While the story may look like two people are having a conversation when the scene cuts back and forth between two close-ups, the real time that has passed between the recording of the close ups could be hours or days. What may have changed is the room tone. Room tone is the "air" that is in the space of the scene. It is what is left when all dialog and movement is removed, and it is what dialog editors use to smooth out mismatched lines or unwanted noises that are recorded during the take. Every time the microphone position changes, the room tone changes.

Mismatched sounding shots occur for many reasons: The position of the lights may have changed, microphones are in different positions, traffic outside may be different, or even insect or bird sounds may have changed. It is our job as dialog editors to make the scene believable by smoothing out any change in room tone.

Split the Scene

I work on one scene at a time and split each region onto its own track, making them look like a checkerboard. We can see in this example that a number of takes for scene 7 have been used and they are not in a consecutive order. The region order for this example is 7C-w, 7-a, 7C-s, 7a-2. For each one of these takes, the room tone sounds very different, and there isn't even a "room" per se since this is an external scene in the desert.

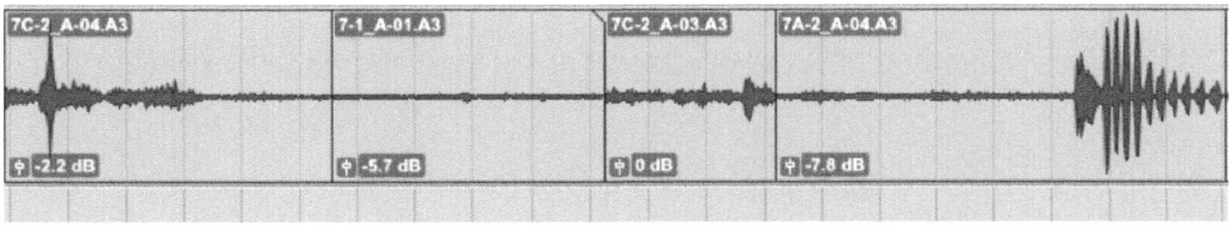

I then move each region onto its own track and extend their handles a little bit past the editor's cut. Next I make a small fade on the beginning and end of each region so that as they overlap, there is a *cross dissolve* of room tone. A cross dissolve, or cross fade, occurs when two audio files overlap each other. During the overlap, the initial audio will diminish in volume while the secondary audio increases in volume. This smoothing out of room tone will soften the abrupt change and if it is done well, the viewer will accept it, especially if there is a shift in camera perspective too.

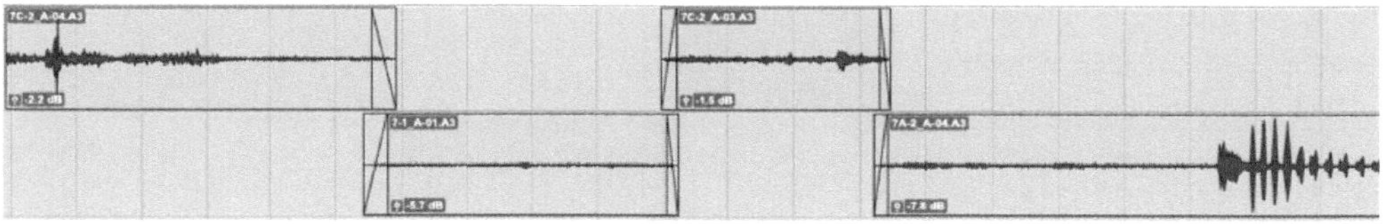

You will find that even if there are a few people speaking and a lot edits, you can keep the track count to around five or six tracks.

Creating Room Tone When There Are No Available Handles

Let's say that there are no handles available and you have to fill in the space with room tone from somewhere else. I find that it is best to look for some usable tone that is very near the space that you are trying to fill. It will be the most consistent since the microphones are still in the same location. When you find a section that is usable, you may have to repeat it or loop it a few times to fill the space. Be very careful that you cannot hear the loop, which happens around half the time that I repeat a small section.

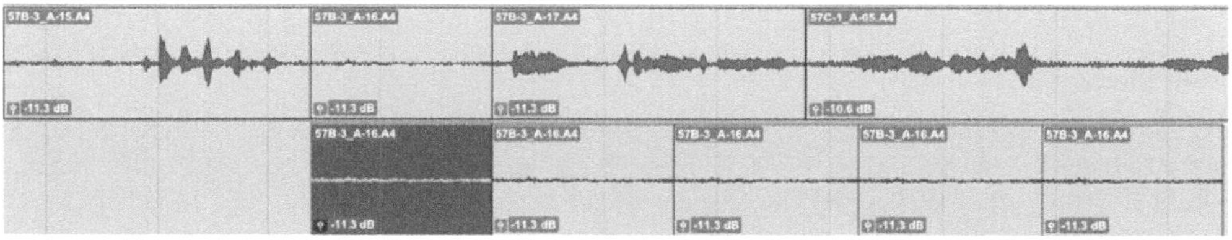

A solution to hearing this loop is to make an extended room tone that is compatible with the scene. I copy the audio to a work track and open the handles to the maximum on either end. In shuffle mode (which will close the gap when I delete something), I then edit out every word, movement, breath, and noise that I hear. What is left is a usable length of room tone that isn't repetitive.

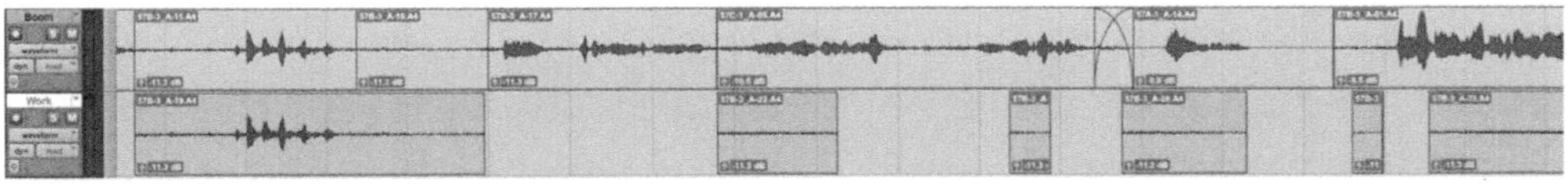

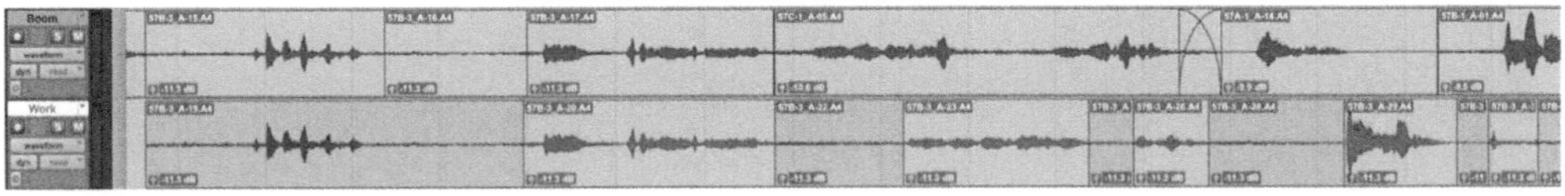

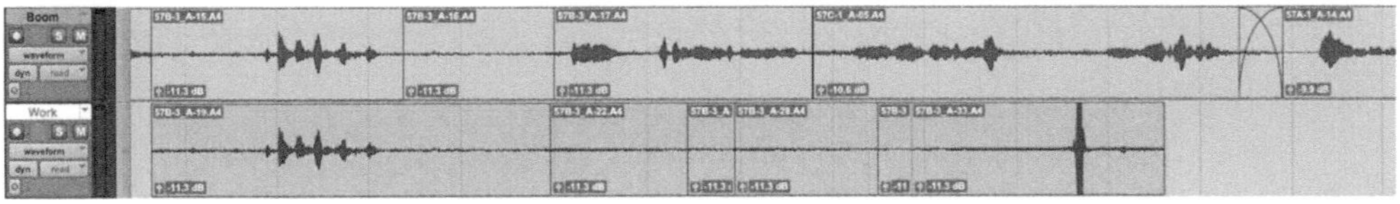

Avoid Adding Supplemental Room Tone

You may think that placing additional room tone will smooth out transitions, but that is not the case. What happens is that adding layers *creates* noise that takes the clarity and punch out of the dialog. The goal is to use only the smallest amount of tone necessary to make the transitions as even as possible.

Remove the Filmmaking from the Film

Your job as a dialog editor is to take the filmmaking out of the film. The audio that is in the OMF is the audio associated with the picture cut, whether the audio is usable or not. On every film I always find many words from the director or assistant director that have been left in the take. Sometimes the picture keeps rolling when the AD gives some direction to the crew and the scene continues. And then there's the director's "action" and "cut" command usually makes it into the audio, not to mention the crew's footsteps. They all have to be removed.

Dolly noises from the camera often leak into the audio and have to be removed by choosing the tracks that have the least amount of noise on them. In some cases, you may have to find an alternate take.

Replace Unusable Sections

For a number of reasons, the audio in the timeline may be unusable. The cause of the problem may be the contact noise of the hidden lavalier microphone under the clothing that is rubbing in the microphone, or it may be a burst of distorted shouting. Often the wireless connection is interrupted momentarily, creating a burst of static. All of these problem audio files need to be replaced, but hopefully you will find another take that will fit into the scene without an edit being heard.

Here is an example from the horror film *Death House*, where a group of prisoners shout "freedom." The take featured in the scene had unusable audio, so I had to find a replacement. It seems that the production sound mixer was taken by surprise for the volume of the group shout, which is understandable. You can see here that the audio was recorded too loud, or over-modulated, or clipped because the top of the waveform is a horizontal, almost like a mountain plateau.

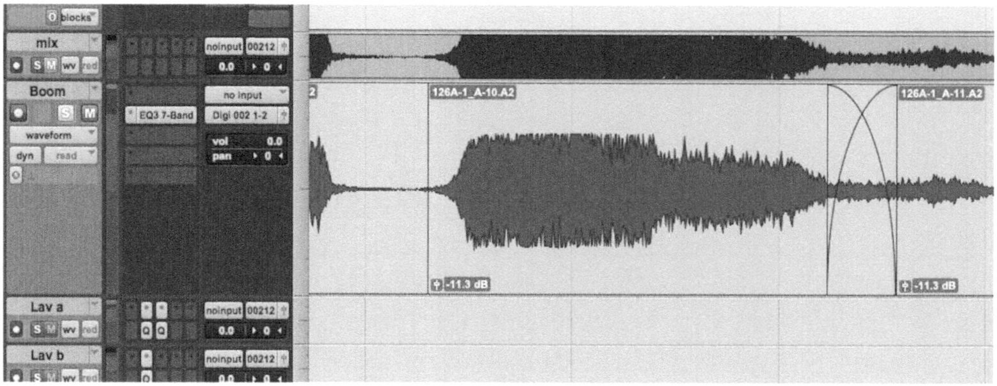

We can also see on the waveform that the audio was from scene 126A. I performed a simple search on the hard drive and I found that scene 126A was recorded on Day 7 and there are several takes. When I listened to them to choose which was best, I noticed that each consecutive take sounded better, probably because the production sound mixer was altering his record levels. I also noticed that scene 126 was on my wild tracks list, and it was in the best interest of the film to check them out. Here there were multiple takes of the shout, and I could hear the director asking for more "giggle" from the actors. By listening to these takes, I found myself on set, hearing what is happening. This was valuable in realizing the directors' vision. When it comes to the final mix, he will remember that there was some insane giggling recorded.

I chose scenes 126, takes 3 and 4, and also the wild tracks. Yes, there are probably more tracks of this shout than I needed, but I could include a bit of one into the surround speakers. This was a decision that I could not have made until I was in the mixing stage of the project, but I wanted to have everything available to me when the time came to mix.

108_T1_4-11009.WAV	25.3 MB
112WILD_T...11010.WAV	13.2 MB
125_T1_4-11025.WAV	50.7 MB
125_T2_4-11026.WAV	48.8 MB
125_T3_4-11027.WAV	63 MB
125B_T1_4-11028.WAV	64.9 MB
125B_T2_4-11029.WAV	90.9 MB
125C_T1_4-11030.WAV	27.2 MB
125C_T2_4-11031.WAV	30.2 MB
126_T1_4-11015.WAV	56.4 MB
126_T2_4-11016.WAV	43.8 MB
126_T3_4-11017.WAV	47.5 MB
126A_T1_4-11018.WAV	57 MB
126A_T2_4-11019.WAV	56 MB
126A_T3_4-11020.WAV	58.7 MB
126A_T4_4-11021.WAV	23.8 MB
126ALT-WI...11014.WAV	11.2 MB
126B_T2_4-11023.WAV	23 MB
126B_T3_4-11024.WAV	21.8 MB
126C_T1_4-11033.WAV	45.8 MB
126D_T1_4-11034.WAV	136.9 MB
126E_T1_4-11035.WAV	68.7 MB
126F_T1_4-11036.WAV	40 MB
126G_T1_4-11037.WAV	77 MB
126WILD_T...11011.WAV	82.6 MB
126WILD_T...11022.WAV	37.5 MB
[illegible]	[illegible]

Part of the sound report of *Death House*

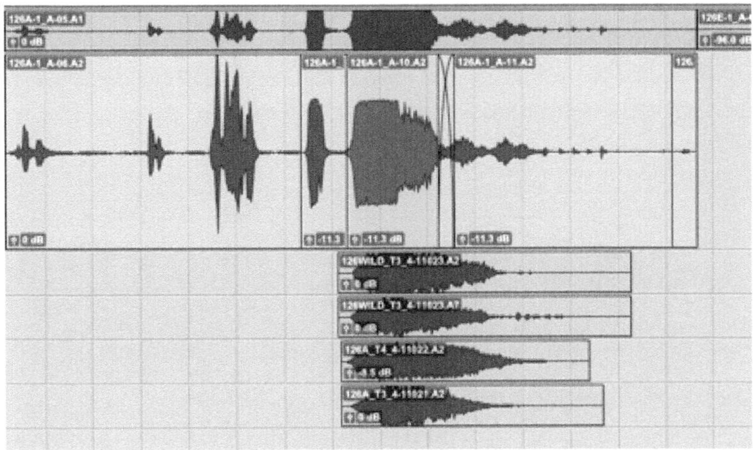

In the same film is an example of a wireless interruption in the dialog. On the second half of the line, "stay by this door, I know this cell," there was so much static from wireless interference that the line was barely audio. And again, it even looked clipped because of the flat top of the waveform. Like the previous example, I searched the production audio files and found another take from scene 118. For the usable take, even though the line was delivered at the same volume, in the same tone, the read was a little quicker than the line used in the shot. Using the production mix (A) as a guide, you can see that by the time we get to the word "door", the transient attack of the consonant "d" from replacement audio (B) was a little early (to the left) of the guide track. By using a time compression/expansion tool (TCE), I was able to pull the audio a little to the right and slow it sown, only about 5 percent, without changing the pitch (C). I did this to all of the microphones in the replacement take, even though I was probably only using the lavalier microphone in the final mix, because I still wanted them in the timeline in case I needed to use the other mics. I checked by ear by listening to the original mix (A) against the replacement audio tracks. It was not an exact match, but it didn't involve a close-up of the actor—the scene is a medium shot in a dark room—I made the decision that it would work and I could move on.

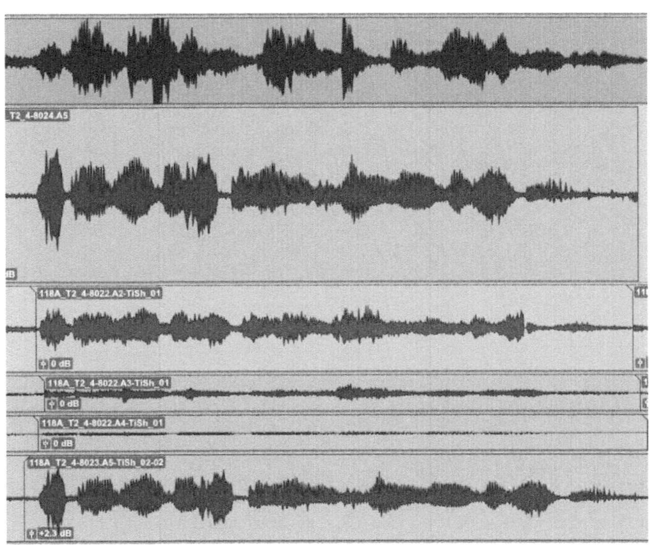

Lining up replacement audio

The time it took to replace and edit the two seconds of dialog took about fifteen minutes. From this example, you can get an idea of how long it might take to complete the dialog edit on the entire feature film.

DECIDE WHAT NEEDS TO BE REPLACED

Your role in the dialog editing process is also to make a list of dialog that cannot be salvaged. This list is important for the discussion that will need to happen with the director in order to plan dialog replacement sessions (ADR). Even if you are the individual that will be recording ADR, having a separate list of lines that need to be replaced will prepare you for the session.

STUFF TO REMEMBER

- Check that the audio syncs with the picture early in the process.
- Go through the film completely as a first pass to assess any problems.
- Make a list of wild tracks that are available to dialog replacement
- Work scene by scene.
- Decide whether to use boompole or lavalier microphones for scenes as appropriate.
- Use room tone as replacement audio to remove unwanted noise from the film.

PUTTING IT INTO PRACTICE

- Create looped room tone from any recorded dialog that you have available.
- Try editing two takes of the same line to make them sound the same.

---

7 EQUALIZATION AND DYNAMICS
Shaping the Sound

There are many tools available to shape the sonic character of audio, and most of them fall into one of two groups. We are either altering the amplitude (or the volume) of the audio, or we are altering the frequency (or timbre) of the audio.

There are many terms for each of these groups.

FREQUENCIES

Let's talk about altering the frequencies first.

Sound is composed of waves of air compression and rarefaction, much like the waves in a body of water. In a lake, the waves are peaks and valleys that ripple across the surface. Sound is very similar in that there are waves of pressure that we perceive, and the symbol for a wave of pressure is a Hertz (Hz). The bottom range of our hearing is around 20 Hz, meaning that there are twenty waves of air compression per second, which is very low. The high frequency that we can hear is around 20,000 Hertz, or 20 kHz. The top of our hearing is dependent on age, and as we get older the highest frequency that we can perceive drops. I am fifty-nine years old and my hearing starts to drop off at around 15 kHz. This is very typical. Try testing your hearing with any number of available apps that will play frequencies through your device—it is no bad thing to know the physical limits of your hearing.

We hear frequencies as pitch and every note that we hear or sing is a number. In Western music, when we hear an orchestra tune to an oboe player, they are tuning to A-440, meaning the A they are tuning to is 440 waves of compression per second.

A tool called an equalizer (or an "EQ") controls the frequencies of audio, either by cutting (lowering) them, or raising (boosting) them. We usually can find EQs in our music playing apps and they look like this.

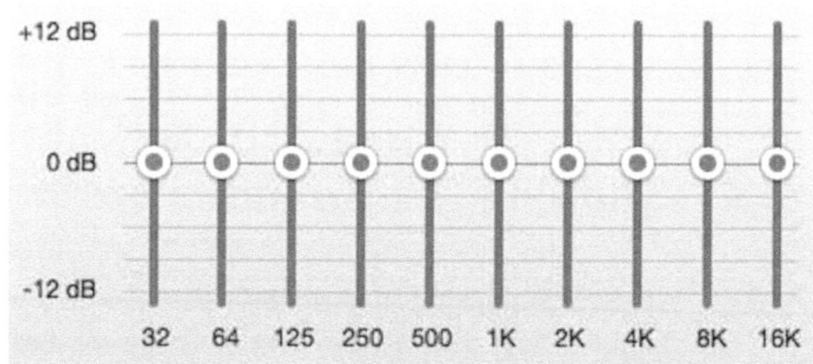

This is a graphic EQ and these sliders have fixed frequencies where we can cut or boost the frequencies, thereby altering the "color" of the sound. I use the word "color" as a way of indicating its timbre. For example, if I say that it sounds "dark," you would think that is sounds muddy and muted, which means that there are many high frequencies. If I say that it sounds "icy cold," you might think that means that it is bright and contains few low frequencies.

On our graphic EQ above, you see the number 500. This means that if we raise or lower the 500 Hz slider, we are turning up or dropping the volume of 500 Hz. This doesn't mean that we are not changing 400 Hz or 501 Hz. There is a curve (called the "Q") around each fixed frequency. So when we increase or decrease the 500 Hz slider, we are also changing the volume of the frequencies around 500 Hz. The width of the Q will determine how many frequencies around 500 Hz will be changed.

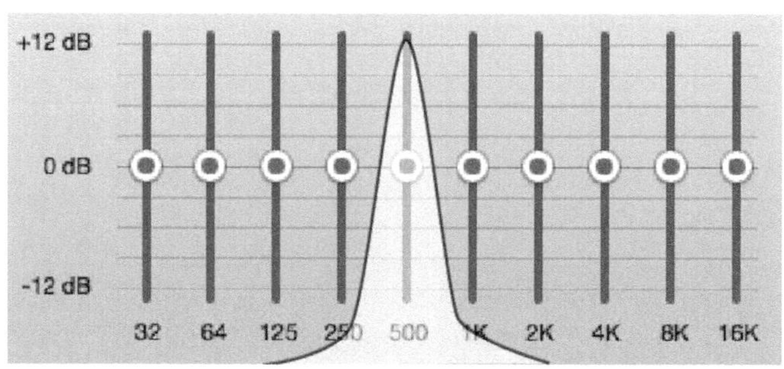

While graphic EQs are useful, they are rather limited in that their frequencies are fixed: You cannot decide which frequency you want to cut or boost, for example. You also cannot adjust the Q specifically and are instead limited to adjusting wide bands of frequencies. Your only choice is the amount of gain that you change.

Parametric EQ

With a parametric EQ, on the other hand, you *do* have the ability to choose the frequency that you wish to cut or boost. In Pro Tools, the parametric EQ that comes with the program looks like this.

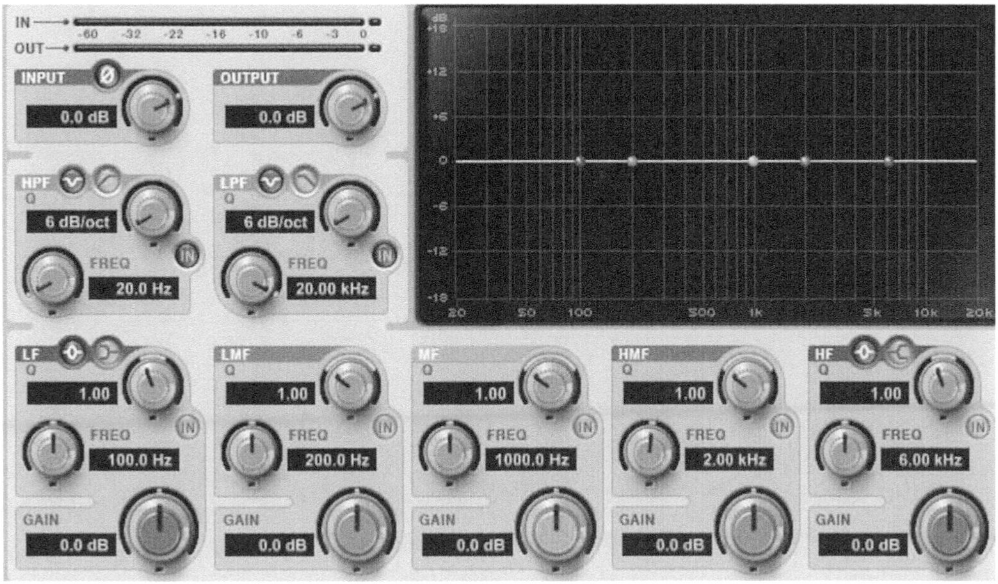

This is a seven-band parametric EQ. The middle five bands are fully parametric and you can adjust frequency (Hz), gain (dB), and bandwidth (Q) on each. I just grab a dot and move it as I listen, then adjust the Q after.

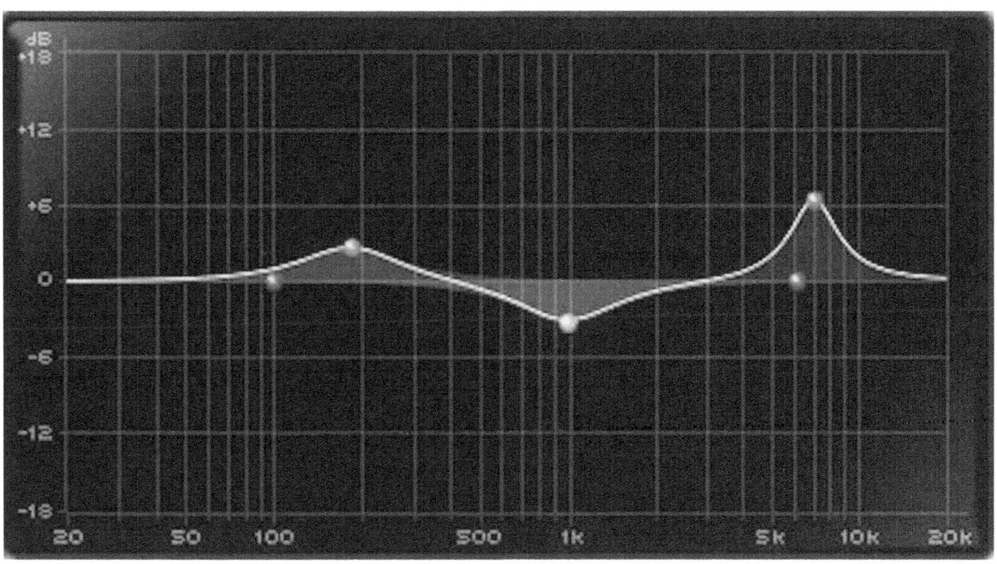

The two bands on the left and right are called low and hi shelf respectively. They can cut or boost the selected frequency and all frequencies above and below the selected number. A hi and low shelf can look like this:

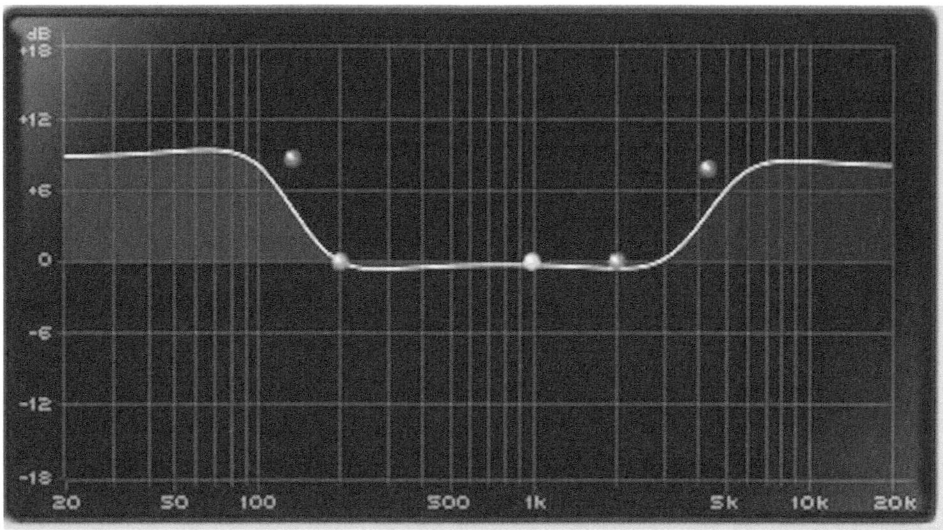

This seven-band band parametric EQ also has two filters: a high-pass, and a low-pass filter.

As noted in Chapter 3, a high-pass filter will pass the high frequencies and filter out the low frequencies, while a low-pass filter does precisely the opposite. While hi and low shelves can drastically reduce frequencies, they will not eliminate them; filters will do away with selected frequencies above and below their filter points, though.

If you activated high- and low-pass filters, it would look like this.

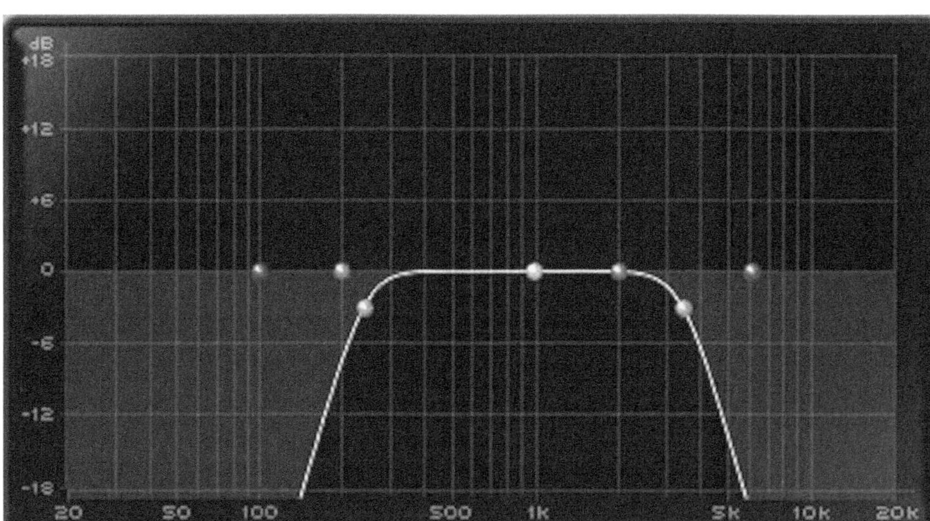

This kind of processing could be called a "band-pass filter" since the only frequencies that we are passing is the band in the middle. This type of filtering is used to simulate the sound of a telephone, since telephones do not reproduce all high and low frequencies.

A word of caution when boosting frequencies in an EQ, however. What you are doing essentially when boosting is increasing the gain of frequencies while simultaneously increasing the gain of the audio track. This may cause the track to over-modulate and clip, creating a nasty sound. I generally use EQs to remove the ugly or unwanted parts of the audio, and if I boost anything, it is usually a little clarity in the high frequencies.

DYNAMICS

Now we are going to alter the level, or volume, or gain of the audio. We measure gain changes by a unit called a dB (decibel, or one-tenth of a bel). A dB is an exponential measurement; for example, boosting the audio by 3 dBs is essentially doubling the volume. *(0 * log10 (2) = 3.01 dB, where 2 is the double in power (volume).)*

The tool that is mostly used to alter dBs is a compressor. A compressor limits the *dynamic range* of the audio, which is the ratio between the quietest and the loudest perceivable sound. For example, a vintage 78 rpm record has very little dynamic range compared to modern vinyl records. I didn't notice how small a dynamic range VHS tapes had until the invention of the DVD and I found myself turning the volume up and down on the television. (Blu-rays have an even greater dynamic range.)

For music-delivery systems, the lowly cassette tape had a narrow dynamic range. When audio CDs were invented, the classical music community embraced them because of their wide dynamic range. The volume of orchestral music can range from a single triangle tap to a full *tutti* orchestra and CDs could reproduce most of this. A wide dynamic range is extremely effective in movie theaters, yet can be very annoying at home since we find ourselves adjusting the volume, but that's a topic for another chapter.

A compressor limits the dynamic range via threshold, ratio, and gain controls. When we set a threshold on the compression, it will only permit the level—that is, the strength of the audio signal—to go past it based on a set ratio. Most compressors default their ratio at 3:1, meaning that for every 3 dB past the threshold, the compressor will only permit 1 dB of level. Here is a well-read, well-recorded narration.

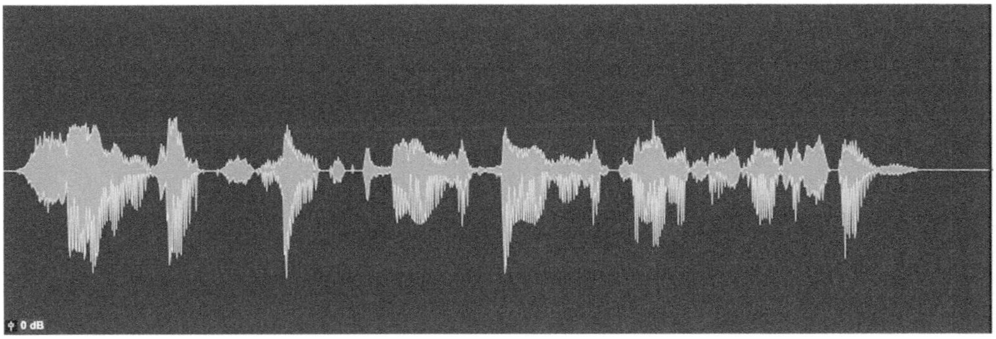

Look at what a 3:1 ratio, with a threshold of 1 20dB can do to narration.

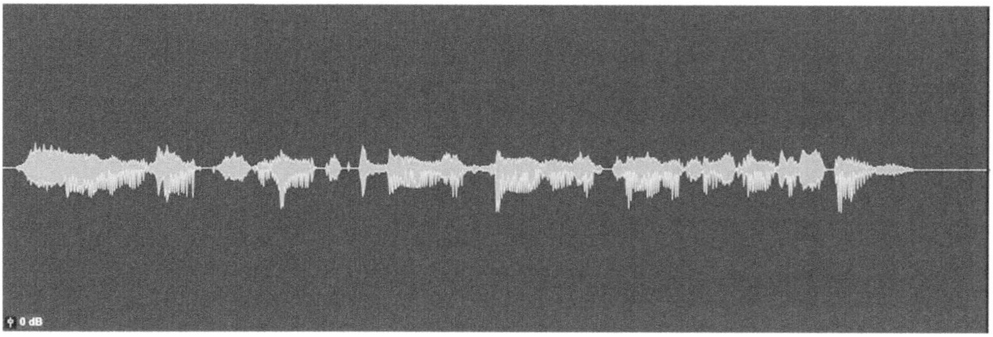

And notice what a greater ratio will do.

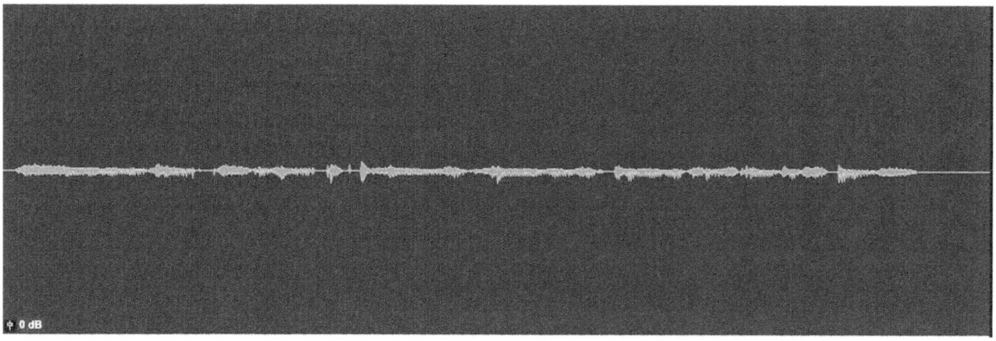

The level of the audio is very low, and we use the gain to bring the entire audio back up to a decent level. In British compressors, the gain is sometimes called "makeup" since it is making up for the lost gain.

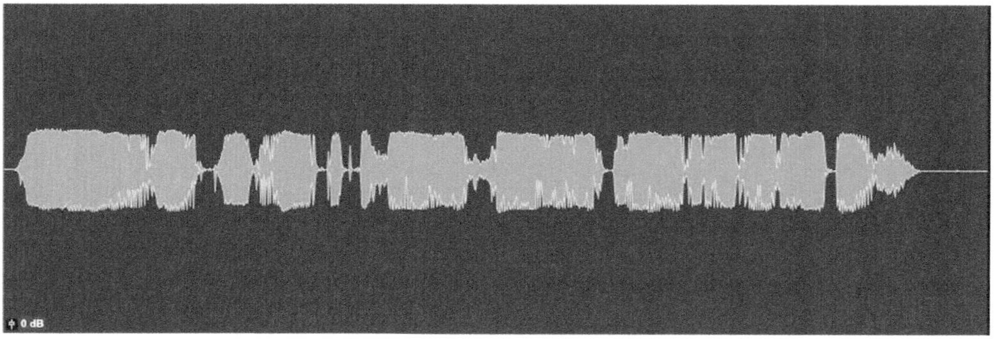

If we look at what a compressor can do to an electric guitar, here is a recording of a single guitar note (an A-220, by the way), recorded directly from the guitar.

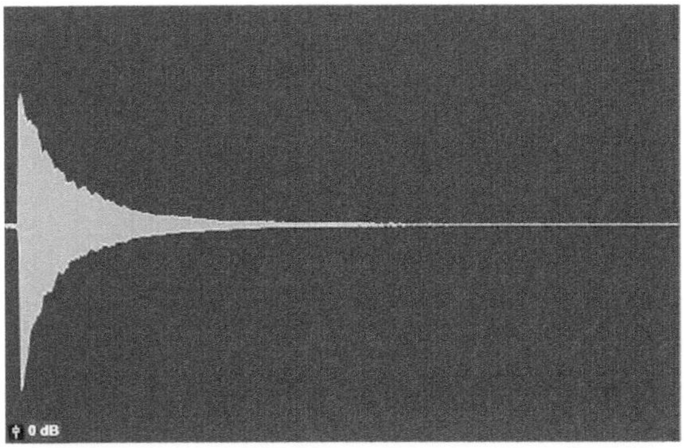

It certainly doesn't sound like rock and roll. There is a peak in the attack, or "attack transient," and then the sound quickly becomes very quiet. If we compress the guitar by a high ratio, and then make up the lost difference in gain, we get a guitar sound that has a very long sustain. Now *that's* rock and roll.

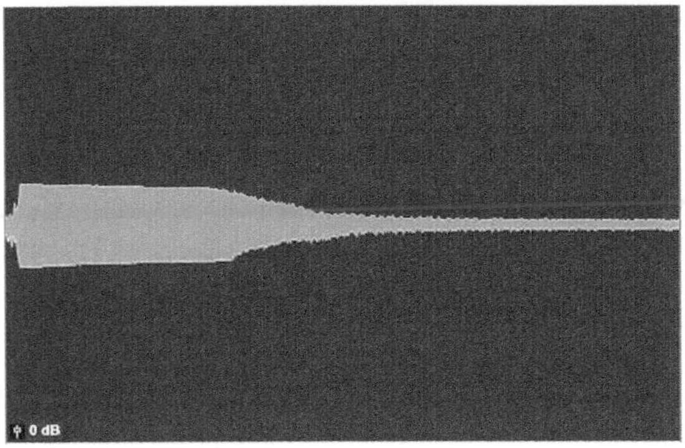

There are other controls on the compressor that adjust the speed of the gain reduction, namely the attack and release. The attack adjust how fast the gain reduction will happen in the beginning of the sound, and the release will adjust the speed at which the compressor will let go of the sound.

On dialog, heavy-handed compression will sound like a radio. The radio stations want every syllable to be heard and they want the sound to project and be noticed. Also, the Federal Communications Commission (FCC) has rules that limit the level being transmitted and compressors do a very good job of this. There is an unwanted side-effect to compression. If a compressor is rapidly attacking and releasing the gain, we hear a "pumping" sound, and whenever the compressor releases the gain reduction, we hear an increase in the background noise, or room tone. When we reduce the level of dialog we are also reducing room tone, and when the compressor releases, the room tone also gets louder.

Loudness Wars

Radio, or whatever radio has evolved into these days, is a marketing tool to get people to buy music, and the louder the music is, the better we think it sounds: Artists want their song to "jump" out of the speaker and there has been an ongoing contest to make the music seem louder. There is a ceiling to volume that cannot be crossed, and so mastering engineers have used compression to squeeze as much volume as possible into a song. Much like the electric guitar, gain is reduced to an even level, and then all of the audio is brought up to the loudest possible level. Since the mid-1990s there has been a trend to make music as loud as possible, sometimes bringing it to the verge of distortion.

Look at this history of songs from the 1960s until today.

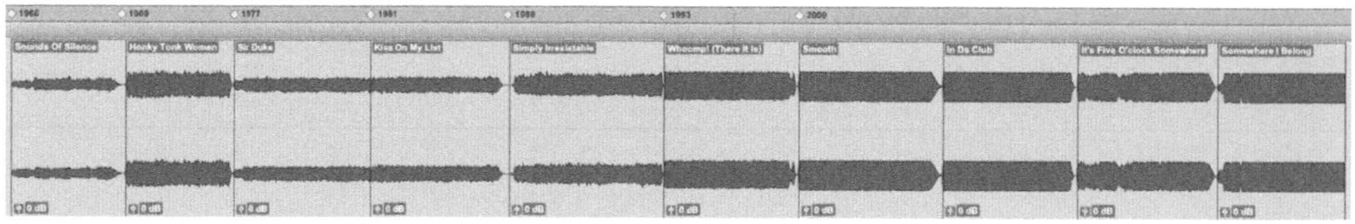

I think that this extreme reduction of dynamic range, along with the level of volume bordering on distortion, creates ear fatigue and removes the emotional content of music. I remember hearing the opening song from *Skyfall* in the movie theater. Adele sounded magical and her singing moved me in a way only music can. When I listened to the song at home on a music-delivery system from my computer, though, all of the life was gone from it and it didn't have the same impact for me. Theaters have wide dynamic range, and music delivery systems do not. Happily, the trend is changing, and musicians are starting to take control of the mastering process.

THE OPPOSITE OF COMPRESSION

The opposite of a compressor is not (as you might expect) a de-compressor, but rather an expander. What is does is set a threshold on the bottom of the level, which when crossed will mute the sound. This is what is found on communication radios such as those used on boats. Boat radios have a squelch control on them, which is an expander. When no one is talking on the channel, we would normally hear static, but when the squelch threshold is set below the incoming signal, the static is muted. The only time that I have used an expander in audio post was during an attempt to eliminate long reverberation in a recording. It is very difficult to get this process right without cutting off the dialog, and indeed it is only rarely successful. Another label for an expander is a "noise gate," as the tool closes the gate on noise when someone is not talking.

De-Essers

De-essers remove excessive sibilance of speech. This sibilance is created by consonants such as "sh," "z," "ch," and "s." If you listen to an unaltered audio of speech and focus on the sibilance, you will notice what I am talking about. Usually the sibilance spike is in the frequency range of 2 kHz to 10 kHz, depending on the individual. De-essers work by putting a compressor on a specific band of frequency. Another word of caution, though; excessive de-essing sounds like you are holding the person's tongue.

Eliminating Hum

In our electricity-driven world, we are surrounded by hums. You may not notice them in everyday life, but if you train yourself to listen for them, you will soon spot that they are everywhere. In fact, they are so ubiquitous that our brains remove them from our consciousness. Microphones however, pick up everything. HVAC, fluorescent lighting, kitchen appliances, computers, and beeps of all kinds surround us and have to be eliminated from recordings. Most of the time they are not that difficult to remove: If you can hum it, that's a frequency, that is a number, and the hum can be eliminated. The first frequency that I go to, without even using my ears, is 120 Hz. Why? Because the fundamental of our electricity is 60 Hz. A *fundamental* is the frequency that we hear as the pitch of a sound, and it carries the most energy. *Harmonics* are frequencies superimposed upon the frequency and which give the sound its character. If I sing a note, and you sing the same note, the reason why our voices sound different, even though we are singing the same pitch, is because we each have unique harmonics. Even ordered harmonics are pitches above the fundamental, usual an octave, then a fifth, then an octave, and then a third, and so on to create an overtone series.

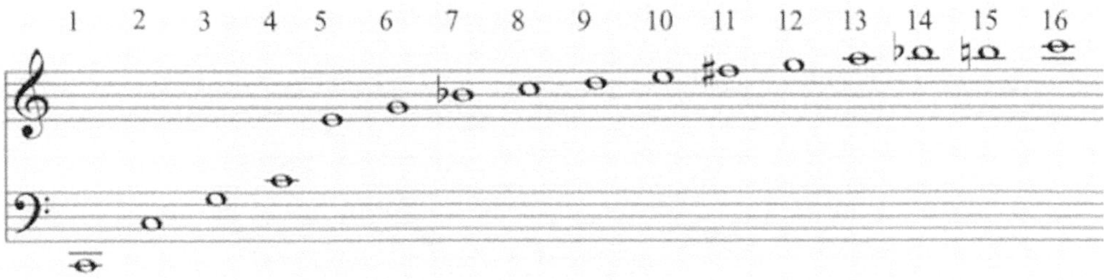

So what does this have to do with hums and electricity? An octave is a double of frequency, so 120 Hz is the first order harmonic of our electricity. Here is a list of the fundamental and the other harmonics of our electricity:

60 Hz *Fundamental*

120 *1st Harmonic*

180 *2nd Harmonic*

240 *3rd Harmonic*

300 *4th Harmonic*

360 *5th Harmonic*

420 *6th Harmonic*

480 *7th Harmonic*

540 *8th Harmonic*

These are the “magic” numbers that will probably remove electricity-based hums. The reason why I use the word magic is that many filmmakers are surprised and delighted when I use an EQ to notch out an annoying hum that is ruining a scene.

A 120 Hz notch, using our EQ, looks like this.

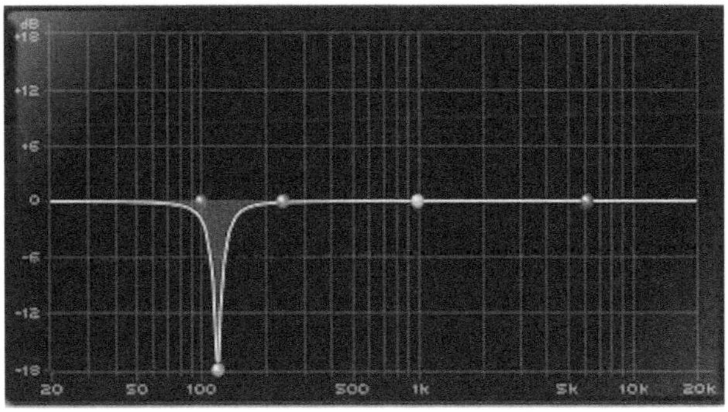

First, I usually boost that frequency to make sure that it is the offending number.

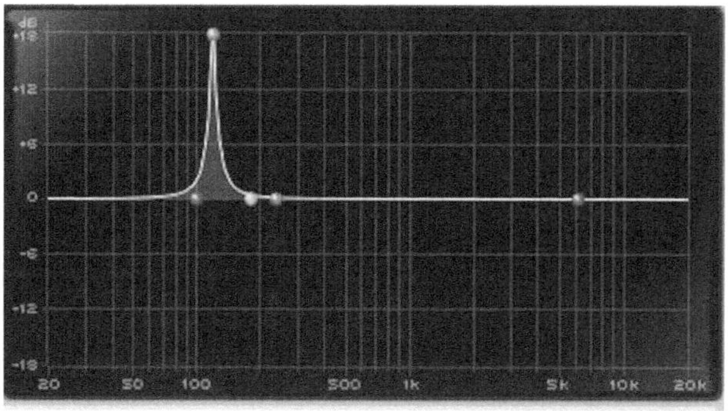

The sound of this is usually enough to clip the output, but I want to be sure that it is the correct number. Removing the first three harmonics of electricity would look like this, but I try to avoid this since it may remove too many low frequencies and make the sound a bit too thin.

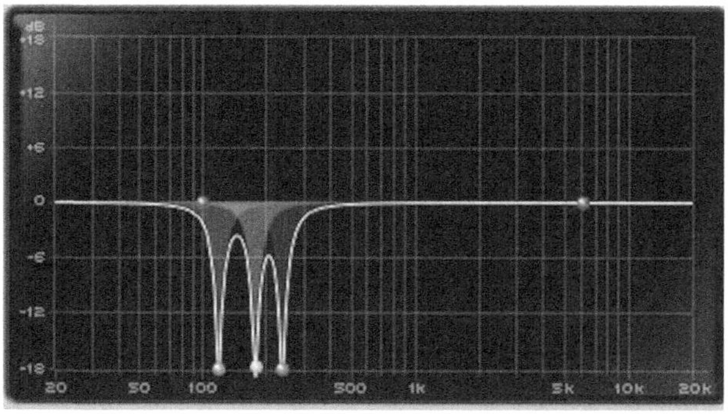

I said above that I would "probably" remove these frequencies first. "Probably" because I have been in some situations where it seemed like the fundamental was 68 Hz, which has me guessing that the electricity may not be right at the location. Use the overtone series numbers as a guide, but always check with your ears.

Resonant Frequencies

There are other frequencies that need to be removed from recordings. These are *resonant frequencies*. Resonant frequencies are natural vibrations of physical objects. The best example is that of a wine glass shattering because an opera singer sang the same pitch as the resonant frequency of the wine glass, which caused it to vibrate so much that it broke. Everything has a resonant frequency: For example cello players know the "wolf" of their instrument, which is the resonant frequency, and have to avoid it if it is not in tune with the scale. Wallboard has a resonant frequency of around 130 Hz, so if we roll off all of the frequencies above that . . .

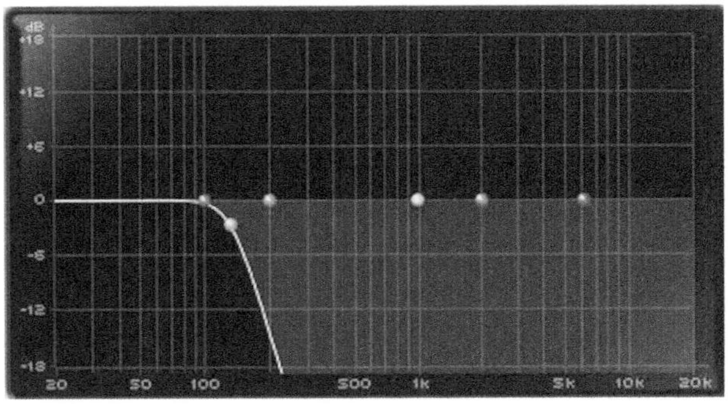

. . . it will sound as though there's a party going on next door.

With this in mind, other situations have to be considered. Let's say there is a scene where two people are having a conversation while seated at a table. The table's resonance will usually come into play. Try boosting a band of EQ and moving it left and right around the low mid-range area, increasing and decreasing the frequency, until you hear an obvious increase in volume. You'll know when it happens. If you then cut that frequency the dialog usually becomes clearer.

Broadband Noise

Broadband noise is a little more difficult to remove with EQ since it straddles all of the frequencies and removing the noise will remove important frequencies of speech, making the sound uncomfortable to the audience and pulling them out of an immersive movie experience. Good examples of broadband noises are a fan, heating and air conditioning ventilation, and the din of city traffic. Hopefully the production sound mixer has done a good job at keeping the noise at a low level. If the difference between the noise level and the signal (dialog) level is small, meaning the *signal to noise ratio* is small, it will be difficult to remove the noise; if the ratio is great, however, it is possible to remove the noise with plugins without greatly affecting the signal.

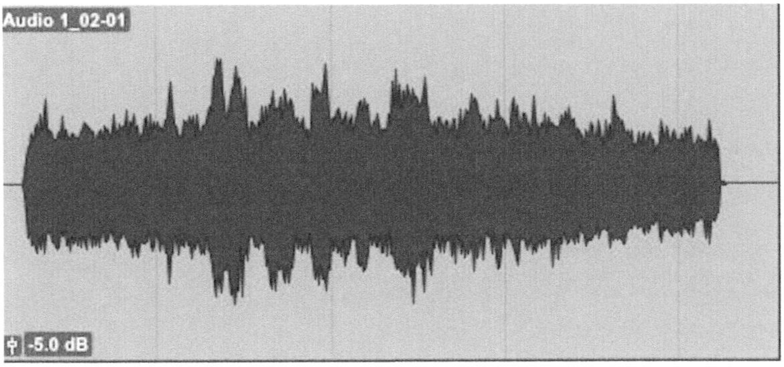

A small signal-to-noise ratio

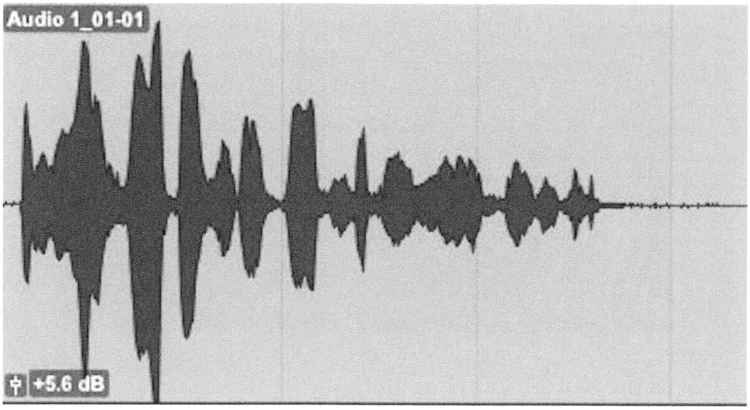

A large signal-to-noise ratio (better)

These plugins work by taking a "noise print" of the offending sound, and removing the noise print when the signal dips below a user-defined threshold. Be careful, though, since the plugins can behind leave unpleasant and unwanted artifacts. After working for years with noise reduction plugins, I can now pick out their overuse on radio and television programming.

A couple of software developers recently marketed a de-reverb plugin that removes room reverberation by using frequency-dependent expanders that gate what is in between syllables. It can work in some situations, and indeed all plugins and filters can help a little, but in reality, the best way to deal with extraneous noise is to avoid it from the beginning.

STUFF TO REMEMBER

- Equalizers cut and boost specific frequencies.
- Parametric equalizers have a curve—the "Q"—to narrow or widen the bandwidth of the selected frequency.
- Compressors reduce dynamic range.
- Heavy compression will increase the level of room tone between words.
- Electrical hums can be diminished by reducing specific frequencies.

PUTTING IT INTO PRACTICE

- Using an EQ, alter some dialog to make it sound like a telephone, and then make sound like it is on the other side of an interior wall.
- Record room tone in a noisy room and try to eliminate specific frequencies with an EQ.
- Apply mild and heavy compression on a narration and decide when you can hear the compression and at what point the change in room tone becomes noticeable.

8 ADR
It's Not Repairing, It's Replacing

You have gone through a few passes of dialog editing. You have cleaned up all of the room-tone inconsistencies between edits. You've scoured through the hard drive to find a cleaner alternative for any damaged audio. However, despite your efforts, there are just some takes that cannot be fixed. It's time for ADR.

ADR, or Automated Dialog Replacement, is the process of rerecording dialog that is not suitable to leave in the film. The process is the modern version of "looping." When film and magnetic audiotape were the formats of movie making, replacing dialog was accomplished by cutting a piece of the film where the audio needed to be replaced and taping it together to form a loop. The loop was played back and projected in a recording studio and actors would watch their scenes over and over until they had the performance and sync memorized enough to record a believable take. When a successful take was recorded, a new loop of film was loaded and the process was repeated.

Just think about how tedious this process was. Can you imagine working on a Fellini film where 100 per cent of the dialog was replaced or looped because the director would verbally direct the actors from behind the camera, making the audio unusable?

ADR can save the movie and help actors' performances, but it's going to take time, and that costs money. The actors have to recreate their performance *exactly* in sync with the picture, while keeping the same emotion. There will be no other actor to interact with during the process and they will be working on a single line, one at a time: Five minutes of dialog replacement may take an hour or two. I do not know anyone who says that they love ADR, especially me. I always ask that the director is present for these sessions, since he or she has the final word on performances. Once the session is done, I want it to stay done, so getting the director to approve all of the takes is imperative.

Making the ADR tracks sound like production tracks is always worth the effort. If a replaced line sounds a little different from the line of dialog that preceded it, the audience will know. Audiences are extremely sensitive to changes in dialog, and even if they do not know what ADR is, they will know that something is not right, and this will distract them from the storytelling of the film. For this reason, if there is a scene with just two people having a dialog, and one only one actor's audio needs to be replaced, I still request that *both* parts be ADR'd. If not, it is going to be obvious that only one person's lines were altered. If both are replaced, they both equally sound good and similar in the quality of the recording, and the audience will never suspect.

I have done many tests with the projects that I work on and the classes that I teach to see if ADR is noticed in an edit. To me, it is always painfully obvious that certain lines are replaced since I recorded the ADR lines, but most of the time my students do not pick up on the replaced tracks. You might think that my sensitivity is a little greater, but I have found that almost anyone is involved in the process of ADR will always know that it is not production sound.

USE SIMILAR EQUIPMENT TO THE ORIGINAL

One of the best ways of matching the sound of the production dialog is to record the ADR with similar equipment. Every microphone has its own sound; every preamplifier sounds unique. It may be in your best interest to check, in advance, with the production sound mixer to see what equipment was used on set. You may not always be able to find out, so I always have a standard setup. On the vast majority of film shoots, dialog is recorded using a combination of a shotgun microphone on a boompole and a lavalier (usually hidden under clothing) connected to a wireless system. For this reason, my ADR recording setup always starts with a shotgun and a lav. I do not place the lav under the clothing because I can always reduce the high frequencies a little to make the mic sound like it is beneath fabric. For the shotgun mic, I look at the scene in question. If the boom operator is doing his or her job well and keeping the microphone as close as possible but just out of frame, I can approximate the same location in the ADR session. Usually it is between four and ten feet from the actor's mouth. Even if I am not certain which microphone was used in the mix, I will always record with two microphones and then decide which one to use during the mix.

Another good way of helping the ADR sound like production sound is to record in a similar location to the one in which the original scene was shot. Is it a large room or an intimate space? Does the room have a hard floor surface, or is it carpeted? Maybe the scene was shot outside: I have recorded ADR outside for this reason, but often I am fighting new problems like birds or urban rumble. Most of the time I can be successful in recreating an outside recording by having the session in a large, fairly "dead-sounding" room.

THE TRADITIONAL ADR PROCESS

I am going to talk about two ways to approach the ADR session. The first way, and the most common one, starts with getting the script ready. Every line that is to be replaced will be on the ADR script, and everyone gets a copy—the actors, the director, the assistant (if present), and you, as the recording engineer. It is possible that some of the lines to be replaced are not exactly the same as the original script: Actors improvise sometimes, or some lines may have been changed on set by the director, and scripts have to be modified. You want the actor to be concerned with acting, so if you see a line on the script and realize it is different in the ADR, take the time to make the changes. Keep your actors in the 'zone'.

Next, the session has to be prepared. Before any line that is to be recorded, I create a "beep track." This involves inserting three beeps, a second apart, that will cue the actor when to start their line. The anticipated (but unheard) fourth beep will be the start of the line. Using a beep track can help an actor prepare, but make sure the level of the beep is reasonable: It needs to be a marker, not an irritant. Also remember that most of the time actors will have headphones on during the recording and loud beeps can hurt their ears.

Here is how the beeps look in the session.

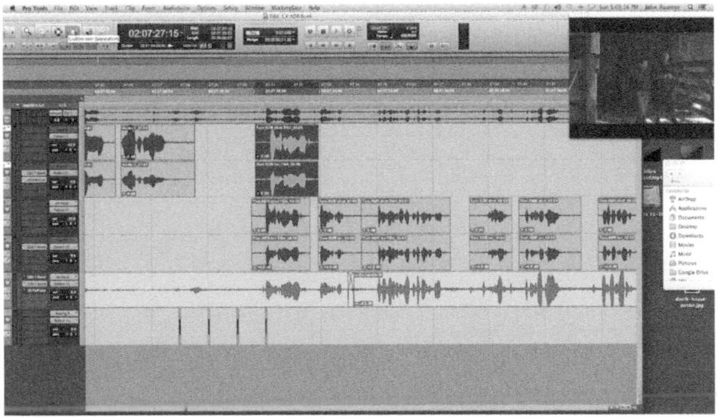

In Pro Tools, ctrl-alt-shift-3 makes a region of selection with 1 kHz at -20db. I made the length of one frame. You'll notice that there is a fourth beep, but it is a muted region and cannot be heard. I use it as a marker so that I can line up the start of the line to be replaced in the appropriate spot. I play the beeps and start recording *before* the actor starts: Recording in advance of the line will give me any breaths associated with the line and plenty of handles that I might need for editing.

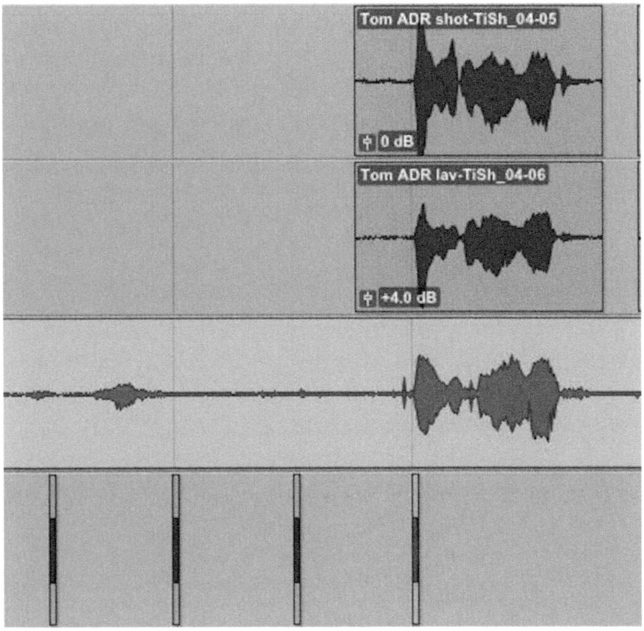

Sometimes the recording takes a couple of passes: The actor may want to hear the original line again, or the director might want to compare a couple of takes. Either way, it is important to stay organized and flexible. For this reason, I use two tracks for the ADR recording: one is the record track, and the other, usually muted, is where I keep alternate takes that are available for playback and discussion. You can see this in the figure below.

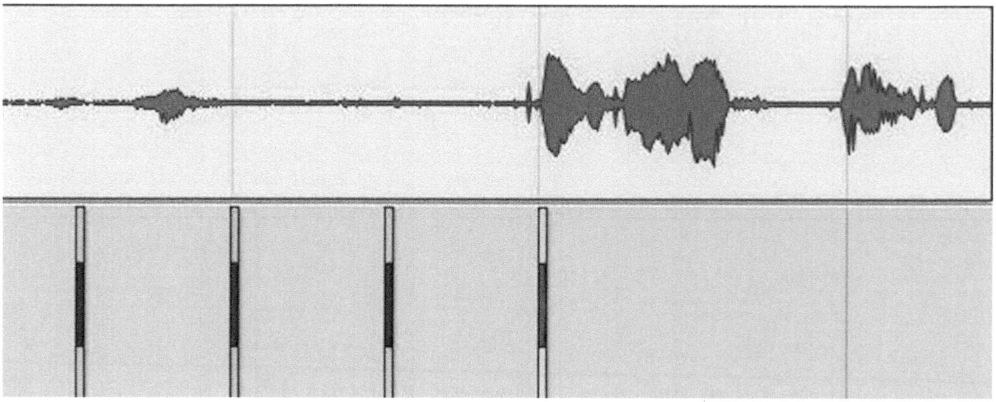

Since you are in charge of the pacing of the session, do not rush yourself, as rushing makes mistakes. An ADR session takes as long as it takes, so prepare yourself for the long run and stay calm.

AN ALTERNATE WAY OF RECORDING ADR

Most of the ADR sessions that I have recorded are for low-budget independent films where directors did not know how bad their production sound was until they got involved in a mixing session. Believe me, it's not a pleasant place to be when they realize that there is a big problem with their "baby" and they ask "why didn't the sound guys tell me this?!". In such situations, the ADR process is not part of making film, but rather damage control and something that was not foreseen. Furthermore, the actors who have worked on these films typically have had little experience with ADR, and the process makes them rather anxious. So it is now my job to put them at ease if we are to successfully get through this and save the film. It is not easy to repeat lines while looking at the scene while staying in sync *and* recreating the same emotion that was in the original. For this reason, I do away with the image, there are no scripts involved, and there are no headphones either, as I am in the room with the actors, usually sitting next to them.

I am looking for them only to mimic the lines when they hear them. I work with small phrases only. What I want them to do is repeat the line right after I play it, so I play it once, twice, and then I say "OK, after this one," and I play it for the third time and right after they hear the line, they repeat it. Ideally they will repeat the timing and pitch of the line, almost like music. Once they get used to the process and learn to trust me, we can accomplish a lot in a short time. The actors lose sight of the broader process and start to zone in on just repeating what they hear. And if what they are hearing is the emotion that the director fell in love with, we are mimicking that same performance. I am aware that this is not a traditional approach to ADR, but in the circumstances of low-budget indie film, I have found that this works.

ALTERING THE ADR RECORDING IN POST: TIME AND SPACE

Very often, the replacement line is not exactly in time with the original, and after you have had some experience working with actors on an ADR session, you will be able to tell when recording more takes will not translate into a better result. Experience will also tell you what you will be able to adjust after the recording session is finished.

You can edit the timing of the ADR lines using the TCE (Time Compression Expansion) tool mentioned earlier in Chapter 6. As discussed, TCE can slow down or speed up the line without changing the pitch but remember that it isn't the answer to everything: Overusing TCE can introduce some odd sounds into the recording, often in the low–mid frequency range. The more you use TCE, the better idea you'll have of what you can change without introducing new problems, and you will also learn what is repairable during the ADR session. And as you get faster, you can make the timing changes quickly during the recording session so that the director will approve the take. This will get the actors to trust you, and be more at ease, and let go a little more as you guide them through the alternate way of recording ADR.

When using TCE, you do not necessarily need to stretch the entire line. Typically most the words match the timing pretty well and only some words or phrases need to be altered. Try to cut up the line in question and match the affected words as far as possible, while leaving the others untouched.

Software companies have done an excellent job in creating tools to help the audio post process. For ADR, one of the best tools is VocAlign by Synchro Arts. This handy tool will analyze the original line, analyze the ADR line, and then rewrite the ADR line to match the timing. This is a powerful software package but again can cause artifacts when overused.

If the timing issue has been resolved with the ADR lines, matching the tonal characteristics of the original is just as important since audiences will note by the shift in sound that the dialog is not "real." There are two areas that need to be addressed: The frequency; and the space.

All audio has a tonal quality defined by frequencies. The characteristics of the sound will have a certain amount of high, mid, or low frequencies. As noted already, our hearing can perceive frequencies from 20 Hz, or 20 waves of compression and rarefaction to 20 kHZ or 20,000 waves of compression and rarefaction. Audio is a representation of sound and every sound has a fingerprint of frequencies that can be scrutinized via a frequency analyzer. There are many available and some are even free.

Here is a frequency analyzer screenshot of a line that needs to be replaced by ADR. It is from the feature *Death House* and the dialog needs to be replaced because it was shot in the shower, with the shower running.

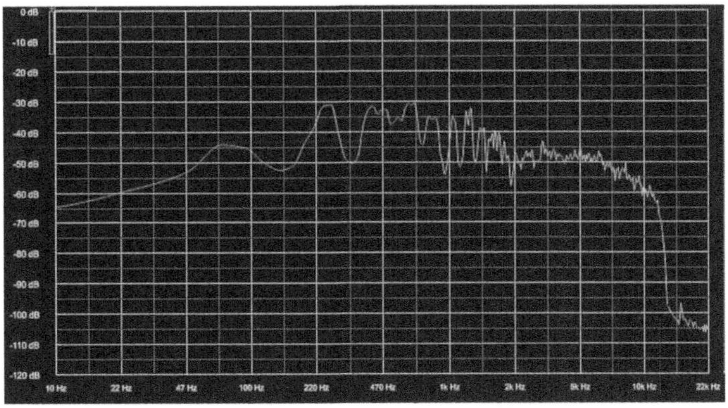

And here is a screenshot of the ADR line.

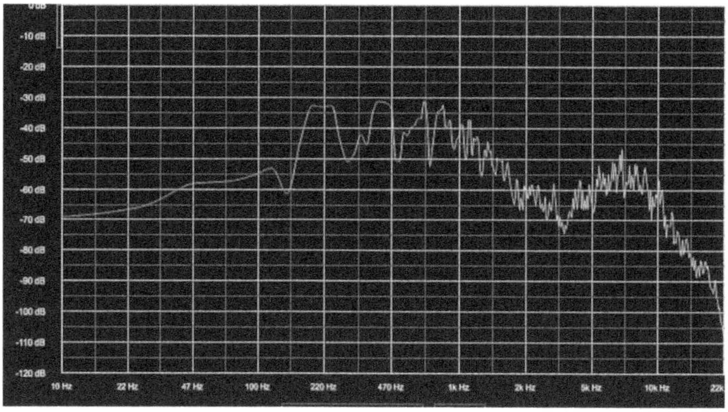

Our goal is to make the ADR line have the same timbre, or frequency characteristics, as the original. You can see that the ADR line has less energy at around 4 kHz, so you would at about 10 db at 4 kHz on an equalizer for the ADR channel. You might want to boost 70 Hz a little and cut 7 kHz some. While this process attempts to make the EQ curve of the ADR similar to the original, it is always best to trust your ears. Do you believe that it is ADR?

Another way to match the ADR recording to the original is to match the space where the original was recorded. This can be accomplished by adding a little reverb (which we will discuss in Chapter 11). For this scene, I added the reverb of a small tiled room to help the ADR sound like it was recorded in a shower.

STUFF TO REMEMBER

- ADR is replacing, not repairing, audio.
- Production dialog is by far the preferred choice of actors and directors and ADR should be used only as a last resort.
- For the replacement dialog to be believable, it has the match the timing *and* the sonic characteristics of the original.

PUTTING IT INTO PRACTICE

- To gain valuable experience at recording ADR and managing an ADR session, replace dialog by recording your own voice. The process will make you better at handling actors and faster at editing ADR. It is a good idea to practice this when talent and directors are not watching you and waiting.

9 SOUND DESIGN

Sound design is the overall aural picture of the film. Distinct from dialog and music, it is brings audiences into the time and space of the story. It creates worlds that would not otherwise exist and recreates atmospheres of the past. Some sound design is narrative in that it mimics what we see on screen, while other types may be subliminal and the audience may not be aware that they are being influenced.

PIECES OF THE PUZZLE

Designing sound for a film is like assembling all of the parts of the puzzle to complete the vision of the filmmaker, so it is very important to understand the filmmaker's needs. To that end a "spotting session" with the director and anyone else involved with the audio post production process is necessary. The person in charge of audio post is called the *supervising sound editor*. This is the person who will interact with the dialog editor, the sound designers, the music editor, and any other appropriate team members. These days, with most individuals working in more that one position, the supervising sound editor could be the dialog editor or the rerecording mixer. In independent film, it could be that there is one individual who is doing it all, which is the situation that I am often in. When I am not worried about deadlines, I find that having a singular vision of the work is pretty efficient. I know what the other parties are bringing to the table because I am the only party and leave space for all of the pieces. Most of the time, however, there is a team, and having the team present for a spotting session with the director is the most efficient way to work. The best way to spot the session is with the OMF from the editor. And the best listening room for the session should be the audio mixing room. Editing bays are usually the worst place to listen to the film. If the music composer cannot be there, someone who knows what concept, or type, of music will be written for the film should be present. For example, if an action scene is going to have a very active percussion-filled score, the sound design team would have a different approach than they might if the score were an ambient music track. Consider this. If all sound effects are loud, and the music score is complex, *and* there is a lot of dialog, what you end up with is just noise.

A careful balance must be achieved to simultaneously engage the listener and move the story along. The same is true for all of the parts of sound design. The parts include ambiences, drones, hard sound effects, and foley, which is another subset that includes footsteps, clothing, body falls and grabs, dishes, money, guns, etc.

LISTENING IN LAYERS

To best understand how to create a soundtrack that the audience perceives as reality, we have to learn how to listen in layers. Creating sonic landscapes involves placing layers of individual sounds together in order to form a single reality.

As an exercise, engage in an intense listening period at place of your choosing. Let's use a typical urban park as an example. There are many individual sounds that make the sonic landscape, from distant backgrounds to up close instances. In the park you will not hear the distant rumble of the city. As you concentrate on closer sound, though, you might hear traffic that surrounds the block and maybe individual cars maneuvering in a symphony of possibilities. The cars brake, pull out, screech, idle, rev engines, beep horns, and start and stop. As your listening focus becomes tighter, you will hear the wind in the trees, some of them in front of you and some as a distant higher frequency static. Then there is wildlife, which includes a diverse variety of birds and insects. Closer perception reveals conversations, footsteps, and incidental human actions like coughing, sneezing, or even skateboards and strollers.

You can pick almost any location and isolate the layers that make up a living soundscape. Even in an interior like a shopping mall, a diverse variety of layers create reality.

Now that you have listen intently and analyzed the world, you can begin to understand the process is creating an artificial reality that, if done correctly, will convince the viewer that they are in that space. When watching really good movies, you forget that you are watching a film and sound design is part of that immersion.

DIEGETIC AND NON-DIEGETIC SOUNDS

It is important to talk about diegetic and non-diegetic sounds. Diegetic or actual sounds are sounds that are in the frame of the film. These are the sounds that we see created on screen, such as dialog, or actions like closing a door. Music can be diegetic if we see the musicians playing instruments, or the music that is coming from a radio, even if the radio is off screen and we understand that is the sound coming from the story world.

Non-diegetic or commentary sound is out of the frame of the picture. If we hear a door outside of the frame that the actors react to, that is non-diegetic. Score music by design is non-diegetic. Sound and music can jump back and forth between diegetic to non-diegetic and this practice is used as a surprise element in horror or comedy.

AMBIENCES

Ambiences, sometimes called backgrounds, are static backgrounds or atmospheres that define the space of the scene. I find that ambiences play a vital role in persuading an audience to believe that they are in the space. When I work on a film, I spend a considerable amount of time editing the dialog so that there is a smooth, clean track of voices. I work to remove any extraneous sounds that would pull the audience away from the story. Once I have removed everything but dialog, I can place the dialog in a background that defines the setting.

On a recent film shot in my hometown of Philadelphia, the director wanted a gritty-sounding film that felt like the city. He spent months scouting locations to get the images that he wanted. When I started collecting ambiences for the film, I went to every location that the scene was shot, at the correct time of day, and recorded a good five minutes of static background. I didn't want any individual elements like conversations or planes overhead. When I placed the ambiences in the film, the cleaned up dialog that I edited now sounded grounded. The added ambiences also helps make the dialog consistent and feel totally unedited.

Ambience location for the film *Crooked & Narrow*, C47 Films, Neal Dhand

How I record the ambiences is a very easy process nowadays thanks to the new, inexpensive portable recorders that are available. I simply use the onboard stereo omni microphones to give me the wide surrounding background, as well as good separation in a surround mix with the dialog in the center channel and the ambience in the left and right speakers. Sometimes I use a third microphone that I plug into the portable recorder and record another channel using a shotgun microphone aiming directly where the camera would frame the shot. This affords me a centered, focused foreground of sound that is in sync with the wide left and right channels recorded with the onboard omni microphones.

HARD SOUND EFFECTS

Sound effects that are in sync with actions on screen are called *hard sound effects*. A door slam, an explosion, a gunshot, or a judge's gavel, are all typical examples of hard sound effects. These types of sounds have to be accurate and come across as realistic or they could pull the audience out of the movie, as any inadequate sound would do. Such effects can be collected together and purchased as *sound effect libraries*, covering a wide range of possibilities or specific genres like automobiles, creatures, or explosions. You can also search online and pay for just one specific sound if you choose.

There are a couple of important considerations with sound effect libraries, however. One is that the best ones are very popular and used worldwide in many circumstances. While you are trying to create something special and different, you could inadvertently use sounds that have been deployed many times over. One of my favorite examples is a servo motor. I have used this sound from a library in a couple of planetarium shows and for years I have been hearing the same sound effect in television and movies too. The next time I have a need for servo motors, I am going to make my own servo motor library by recording a variety of little toys and electric motors. This is something that I would encourage anyone involved in audio post production or sound design to do when they are involved in a project and there is a need for a specific type of hard sound effects that will be needed frequently and in a variety of situations.

Another consideration is that sound effect libraries are dry. In this context, *dry* refers to a lack of any reverberation. If these dry effects are used in a reverberant space, like a cave, they are going to sound as though they do not belong and should not be part of the scene. While this issue needs to be addressed in mixing, the dryness of these sounds allows us to manipulate them when necessary via some of the following tools:

- **TCE.** Often, the timing of hard sound effects from libraries does not quite fit the timing of the scene and they need to be shortened or lengthened. Take a sliding door, for example: The library effect may be too fast for the picture. We've discussed the Time Compression Expansion tool in previous chapters, and as discussed it allows you to "pull" sound to make it longer or shorter as needed. All you have to do is drag the waveform until it fits the picture. Changing the timing via a TCE will not alter the pitch of the sound but overdoing it will introduce unwanted artifacts that may make the sound unusable.

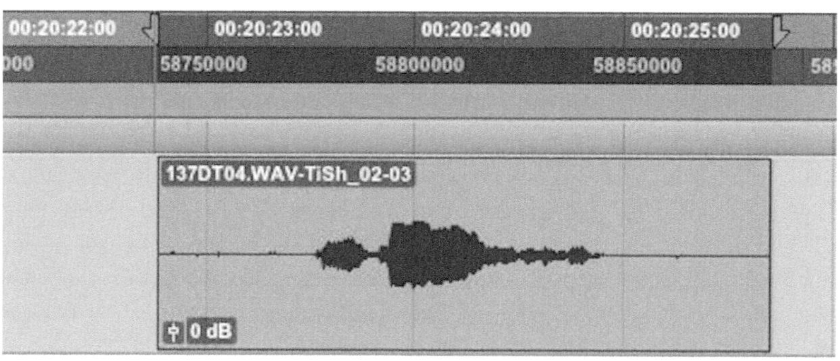

- **Pitch.** Repetition of the same sound effect will grate on the audience's hearing and should be avoided. If I need to use a sound more than once during a short time frame, I may pitch it one up or down a little to make it sound different. A *pitch shift* tool allows you to change the pitch of audio without changing the length of the audio. As with TCE, overusing pitch shift will create unwanted artifacts.
- **Reverse.** Reversing the waveform essentially rewrites the audio backwards. I reverse waveforms to create "other-worldly" sounds or perhaps to indicate entering or exiting a flashback sequence.

Layers

If you are trying to recreate reality by using sound effects from libraries, most of the time they will not reinvent what we hear in real life. Go back to what you have learned about listening in layers. Take a punch, for example. There are many little sound elements that make the effect, such as the snap of clothing, the whoosh through the air, the slap of impact, the crack of bones, and maybe some squish of blood. Making realistic-sounding hard sound effects involves combining more than one sound, then, and searching through library sound takes time. Also you won't know which sounds are going to work until you try them in your timeline. Should the new sound be quieter or louder? Maybe it should be pitched down a little. Or it might sit in the layer better if it was a little faster. Any one, or all, of these choices may work.

FOLEY

The word "foley" came from a guy named Jack Foley who started the process when "talkies" became popular in the early 1930s. Since microphones in those days could only pick up dialog, Foley was employed to fill in the other sound. He and his team did this by projecting the film and recording the sounds, in sync, onto a single track, in real time. A lot of his techniques are still used today.

Foley is the rerecording of all of the human movement in the film. These include, but are not limited to, footsteps, clothing, punches, kisses, eating, cooking, cleaning, or basket weaving.

In my view, there are two reasons why foley has become a necessity in movies and television today.

1. The goal of modern production sound recording is to record consistent dialog, even at the expense of any other sound that is needed to tell the story. Production sound mixers will strive to eliminate any other sounds like insects, the wind in the trees, or footsteps. In wide shots, the only useable microphones are probably wireless lavaliers that are hidden under clothing and those mics will not pick up footsteps. Good production sound recording is sterile and foley is needed to bring the viewer into the space of the film. It grounds the viewer into the scene.
2. Most deliverable requirements for distributors include fully filled-in M&E mixes. M&E mixes are music- and sound-effects-only mixes that are used for foreign-language versions of the film. The distributor wants to drop in new dialog in another language and the film will sound like the original English mix with little effort, and cost, in further intervention. In making an M&E mix, the rerecording mixer will have to mute all dialog tracks, which means and human movement sounds will also be muted.

The author's home foley setup

For these two reasons, I will fill in foley for the entire film, whether scenes need it or not. I record a complete "footsteps" pass of the whole movie, and then do the same with a clothing pass. Usually when I am working on either the clothing or the footsteps, I am also filling in other foley aspects such as opening doors, which would happen in a footsteps pass. On a clothing pass, I would also record, for example, a mattress or frame squeak to accompany someone getting in or out of bed.

Recording Footsteps

Much has been written about foley footsteps. Jack Foley was able to record a couple of people walking, in real time, by using a walking stick. Foley pits are now built into recording studios and have a variety of surfaces including concrete, gravel, sand, wood, slate, etc. I have recorded footsteps many times on these surfaces in quiet rooms, and I have also built some of these flooring surfaces to use in teaching. In reality, I believe the most efficient way—and also the most *realistic* way—to record footsteps is to take a walk, literally.

For this, I spot the film and note which scenes are interior and which exterior. Armed with a shotgun microphone in one hand and an iPad or any other device playing the film in the other, I go to a similar location to where the scene was shot and while pointing the shotgun mic at my feet, I walk like the people in the scene. This way of recording yields a much more authentic sound. Depending on the scene, I may also strap a lavalier mic on my ankles so that I get some extra "grit." If the scene is in an interior, I choose a very similar room and surface to walk.

Recording Clothing

For clothing foley, I put together a setup in front of my computer, which will allow me to record clothing movements for the entire film. With my computer in front of me on a stand-up desk, and wearing a set of headphones, I keep a variety of clothing within easy reach cover my needs. It doesn't take a full closet, I hasten to add: Usually an overcoat, a leather jacket, and three or four different-sounding fabrics will do the trick. I record with a shotgun mic kept close to me as the primary track. I also use two other microphones that I will choose as a secondary mic to record on an adjacent track. One is a hypercardiod that is placed about two feet away that I will use for some space or room. If I'm recording something that is a little loud (like a fight) using a leather jacket and I need some distance so that the recording doesn't clip, I make sure there's lots of space around me. As an option for my secondary mic, I place an omni mic at the top of my studio and about ten feet away. I may need some room to the recording and having a mic recording the room helps the mix. I may or may not use the track, but at least I have it as an option when mixing. And no reverb plugins sound as good as a real room.

You may be thinking that if I am looking for a reverberant-sounding room for clothing or any other foley, why don't I try recording in stereo since that would create a better sense of space? I had thought that too. However, when I moved on to a mix, the foley never felt "married" to the dialog, especially in a surround mix. Dialog comes from the center channel in a 5.1 surround mix, and your newly recorded stereo foley would be coming out of the left and right channels. The perception is that they are not genuinely matched, and in fact that is true: The foley is doing the opposite of what it is supposed to do by grounding the dialog into the scene. In short, foley is mono.

I find the process of recording foley a lot of fun. I get to move and act like the actors and the reward is huge in making the scene believable. All of the subtle little sounds are crisp, clean little gems that are an easy fit when we get to the mixing process.

WALLA

Group voices that crop up in the background are called "walla." We typically hear walla in places like restaurants and at sporting events. Legend has it that the reason for its name is that actors in American English used to mutter "walla, walla" to give the impression of crowd noise. In UK English, it's "rhubarb, rhubarb", and I am sure every other language has it's own version. While sound effects libraries do an excellent job at providing very useful walla backgrounds, I have found myself in situations where I need specific walla that I have to record. For example, I once needed a specific protest chant, so recorded a group of people shouting exactly what I required. You have to be careful of the guy with the big mouth who wants to stand out (and he usually does), as if he goes too far the recording would be unusable.

WORLDIZING

Worldizing is a sound concept created by Walter Murch back in 1973. The idea is that if we as sound designers create a soundscape of a particular scene and the camera is moving in that scene, we would produce the same sonic perspective as the camera movement. Worldizing is done by playing back the created sound through amplified speakers in real-world environments, and then moving the boompole microphone through the space in a similar manner as the camera. It is a simple concept but makes a lot of sense in creating realistic soundscapes.

I once had to recreate a party scene where the actors were dancing to no music, which is the correct way to shoot the scene since the background music had yet to be decided and licensed. When a song was eventually chosen, I couldn't just drop it into timeline of the mix: It would sound too sterile. So what I did was to play back the song through my living room stereo. I then placed a pair of omni microphones in the room and recorded the song in that context. The result was a realistic diegetic-sounding music track that matched the picture.

On another project, using a similar scenario of placing a song in a room, the camera was moving through the room. To create a sense of space based on the camera movement, I recorded a third track by carrying a shotgun mic on a boompole and moving through the room in the exact way that the camera moved through the room.

On the film *Death House*, I needed to create the sound of a constant prison riot that was heard for a considerable length of time (thirty minutes) and from a variety of locations: In the same room, or a floor above us, or even from inside an elevator. I had some appropriate sounds from production recordings and pieced together about three minutes of "rioting." I then took the recording to the abandoned prison where the film was shot. I played the edited riot recording through a speaker and placed five microphones around the prison to capture the different locations in which the rioting would have happened. By having five simultaneous tracks, I could switch the tracks as the camera changed locations.

STUFF TO REMEMBER

- Sounds are either in the world of the picture (diegetic) or outside it (non-diegetic).
- Sound design is created in layers.
- You can alter each layer through a number of audio processors.
- Worldizing records the camera perspective of studio-created sound design.

PUTTING IT INTO PRACTICE

- Practice replacing footsteps by mimicking the same moments from a film. Use your smartphone.
- Repeat the process with clothing.
- Create impact sounds by manipulating your own recordings. Use pitch change and reverbs.
- Create your own personal library of ambiences, including exteriors during the day and night, urban and rural.

10 EDITING MUSIC The Soul of Film

THE POWER OF MUSIC

When a movie is done right, it so believable to the audience that they will forget they are watching make-believe. They should feel like they are inside the story. When the credits roll on a movie we enjoyed, we are sad that the dream is over. Yes, movies can be like dreams.

In making films, an enormous amount of effort goes into creating this artificial reality from nothing. First, the story must have a believable plot, and the actors must be convincing. Sets are created, whether real or computer-generated, that will recreate history or project us into a time far into the future. We have to hear every word clearly, and great sound design will help put us into the environment. Appropriate clothing is created or acquired for every member of the cast. An army of professionals have created a believable world where, let's says, the luxury liner is sinking in the North Atlantic. As the ship is struggling through the waves, we hear an orchestra playing music! Where did the orchestra come from? It doesn't matter: Without the orchestral music, the boat just sinks. With the music, the audience is drawn into the tragedy and feels the catastrophe.

Music is arguably the most powerful way of communicating emotions. I have been to art museums and have never seen someone standing in front of a painting crying. Yet, every one of us knows a song that renders us speechless when we listen to it. That's powerful.

TEMP SCORES

In the last ten or fifteen years, a trend has developed whereby picture editing always happens with some kind of music in place. Previously, this was very rare: Composers would write a score to an edited picture, and the picture would work without the music. Added music would enhance the existing picture. Nowadays, though, "temp music" helps make the picture decisions. I am not convinced that this is the best way, but I have strong opinions on this subject since I am also a film scorer.

Temp music is pre-existing music that is placed in a timeline, usually in a rough cut, in order to help editorial decisions regarding mood, emotions, and pacing. There are good and bad reasons for using temp scores when working with a composer. One good reason is to direct the composer. Given a temped scene, a composer can see the intent of how the scene might feel. If the temp music is liked by the director, it could give idea of the concept of the music that is wanted, which could be electronic, rhythmic, orchestral, melodic, ambience, etc.

The problem with temp scores is that once the music is played against picture, the decision has now been made about what the scene will sound like and it is very difficult to let go of the temp. The director may play the film with the temp track many times, so much so, that changing the music becomes difficult: I call this "demo love." It is possible that the director will play that scene dozens of times with the temp and very few forget what was there and remain open to a new, custom approach to the score.

While higher-budget movies have music editors or music supervisors to edit in temp music to a film, very often I find that the picture editor is the person selecting and editing temp music. This makes him or her part of the creative process of directing the score, and their primary concern is with picture. Sometimes, especially in non-fiction storytelling, the temp music will stay in the final cut of the film.

In this chapter, I would like to discuss two scenarios with mixing score music. One focuses on using library music selected by the director and editor, and the other on original music and working with the composer.

WORKING WITH LIBRARIES

Music libraries are now extremely popular: They sound great and their accessibility by the Internet has made searching by mood or genre easy. Just type in a few words and you can preview music before downloading your choice. This comes as a stereo mixed track, meaning that the mix you hear is the mix you get. The only decision that the editor needs to make is how much or how little will be used in the timeline. The editor may have to edit the track for length, sometimes repeating section of the music to extend it. Since the editor is primarily concerned with picture editing and storytelling, some of their music edits may have to be improved upon. Or, you may be the individual who is tasked to edit music.

The right approach to editing existing stereo music tracks depends on the type of music. With orchestral music that is of a classical nature, the edit is fairly easy, and uses a lot of crossfades. If the story needs the music to make an impact, or something changes in the story where music is needed to magnify that change, a point in the music is chosen and parts of the same track can be edited to ensure it is of the required duration. Crossfades are then used to smooth the edits. This is a fairly simple process, and a good reason why classical orchestra music is so effective in film scores. The music can change in an instant to follow a specific emotion at a specific instance.

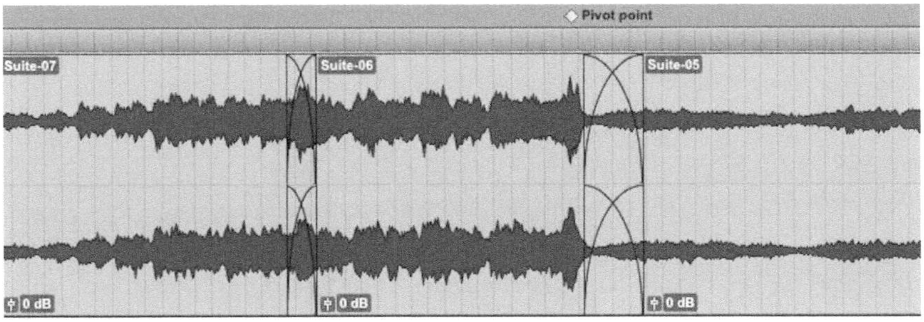

With contemporary music that has a rhythm, cutting the track and attaching it to another part of the track is not a simple process, though. There are musical rules that cannot be broken. Once a beat is established, you can't change the consistent rhythm or the audience will notice the edit and the music will have lost its flow. It will detract from the movie experience. Someone might fall off their chair.

Contemporary music before the mid 1980s had fluctuating tempos. The band would slow down and speed up depending on the feel of the song. If you listen to the early Rolling Stones or Led Zeppelin as a couple of examples, you may notice some parts of the song are a little faster than others. Typically the chorus of the song is a little quicker and has more energy. What happened in the mid 80s is that songs were produced with a "click" track, that kept the band's tempo consistent throughout. The vast majority of music produced since that time has been driven a click track and the tempo is the same from the beginning until the end of the song.

Just about anybody can count along with a song. 1, 2, 3, 4. Count along with your library track and find the downbeat, or the "1." Find a few of them.

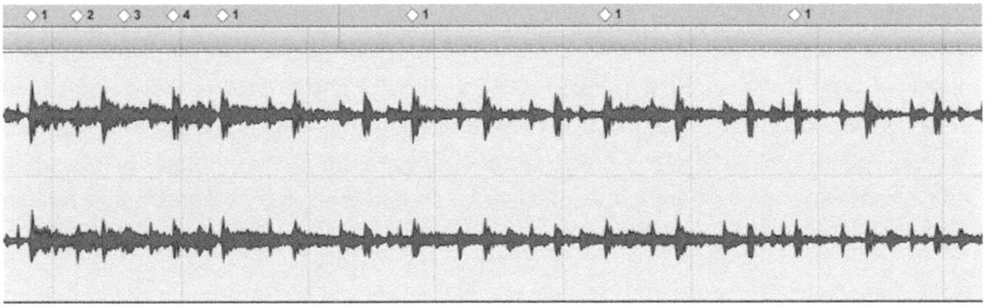

Once they are established, you can cut the track on the downbeat and attach it. Think of it as blocks on a grid.

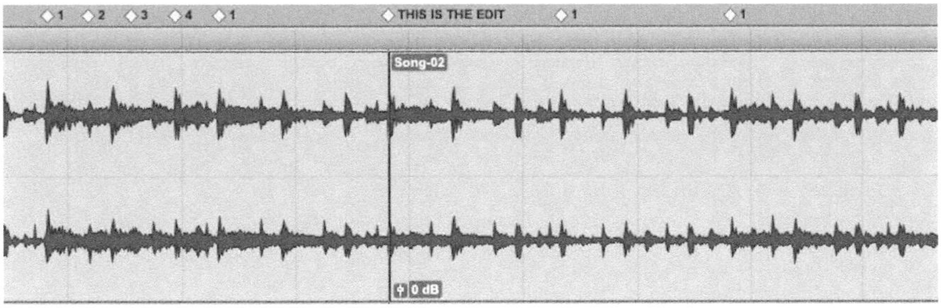

When you have two edited parts of the track locked in place at the downbeat, you can drag the edit anywhere, not even on a beat, to the left or right of the original edit and the track will still stay in time because the track was created with a machine-generated click.

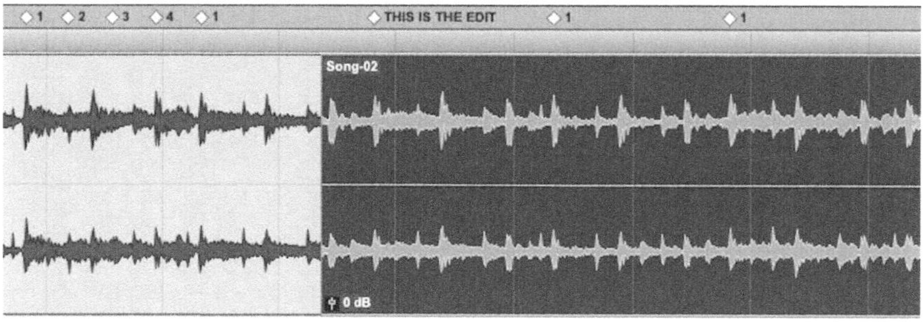

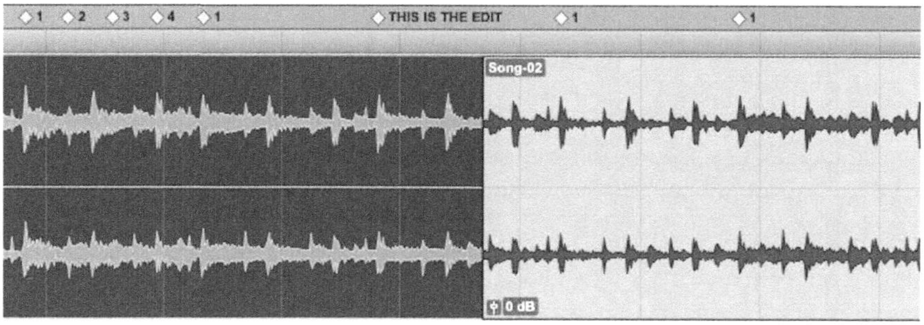

WORKING WITH COMPOSERS

When a composer is hired to write music for a film, each instance of score in the film is called a *cue*. The composer will have had a spotting session with the director and decisions will have been made as to when each cue would start or stop. He or she delivers a perfect custom-fitted score and that should be that. However, sometimes the timing of the picture changes after the composer delivers but before the final mix is complete. In these situations, the audio post person may be called upon to adjust the score to accommodate the picture changes.

Going about editing a custom score is the same process as editing a stereo track from library music. You may have an added advantage is the score is delivered as *stems*.

STEMS FROM COMPOSERS

Stems are the individual parts of the score that, when combined, sound the way the composer intended. They may be as few as four stereo tracks, or as many as twenty: It all depends on the style of music and how the composer has decided to deliver. Before the composer hands over the score to you, it may be wise to have a conversation with him or her to discuss mutual expectations.

Stem tracks that are few in number may contain the following:

1. Rhythm (drums, percussion);
2. Keyboard (piano, organ, etc.);

3. Bass (bass guitar, low pulses, anything in the low range of frequencies);

4. Pads (sustained instruments, keyboards, or synthesizers playing chord parts).

I prefer (and deliver) every instrument on its own track. However, I make an effort to keep a track count to a minimum. If an instrument—let's say an oboe—appears every now and then, I won't keep its track empty for long periods of time. I will try and use the track with other similar instruments. For example, I might combine the oboe with a solo acoustic instrument. Keyboards can be combined if they do not play at the same time, as can drums.

Here are the music tracks to the horror film, *Death House*.

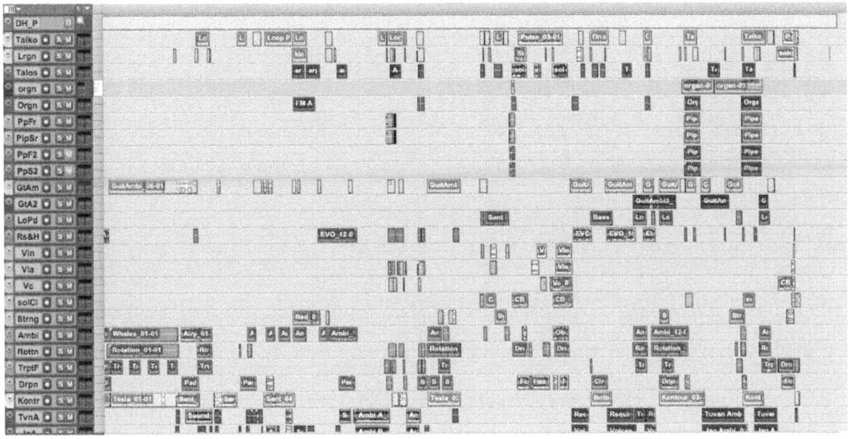

There are twenty-five stereo tracks where every instrument is on a separate track. Stems like this have a great amount of flexibility when editing (and mixing, by the way). With multiple tracks, you may be able to repeat a single track to serve the purpose.

If the director wanted to extend the music for a few seconds into the next scene, you could repeat the ending of the music a couple of times. That may be difficult with a stereo mix because you are repeating all of the instruments and it may not lend itself to repeating. There may also be rhythmic problems, as discussed above. With multiple tracks, however, if you needed to extend the track to the marker ("EXTEND") in the track below . . .

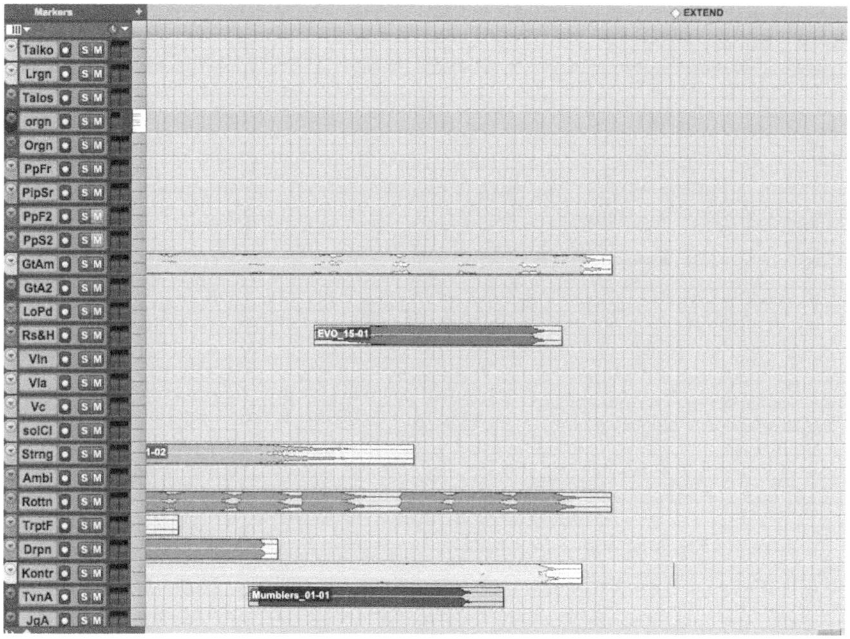

In this example, there is a track called "Tesla" which is a sustained sound. It is easily repeated in the edit without sounding erroneous, and the cue has extended into the next scene as intended.

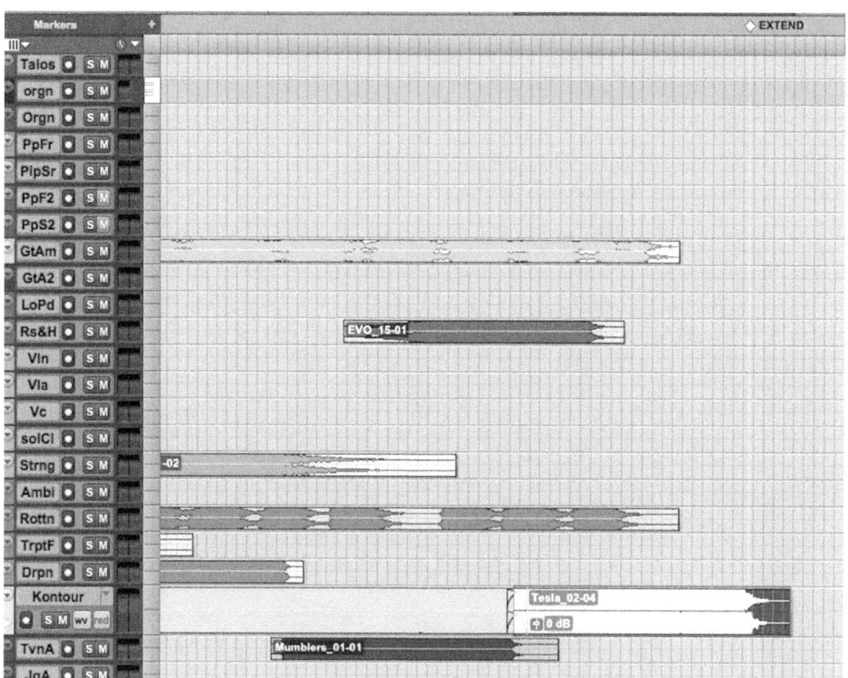

MUSIC CUE SHEETS

A music cue sheet keeps track of all the music that is used in a film or television episode. Composers are paid royalties when their compositions are used and without cue sheets it would be impossible to monitor their use. It is a required deliverable. Cue sheets contain the placement of the music (with timecodes) as well as information on how it is used (background, theme, etc.), and who wrote it. Since you might have a conversation with the composer when you discuss delivery of the stems, a dialog about a cue sheet would be a good idea.

Cue sheets are used by PROs (Performing Rights Organizations) to collect royalties for composers. There are three PROs in the United States—the American Association of Composers, Authors, and Publishers (ASCAP), Broadcast Music, Inc. (BMI), and SESAC—and every country has its own. Here is an example of a cue sheet from BMI, but most cue sheets are similar to this.

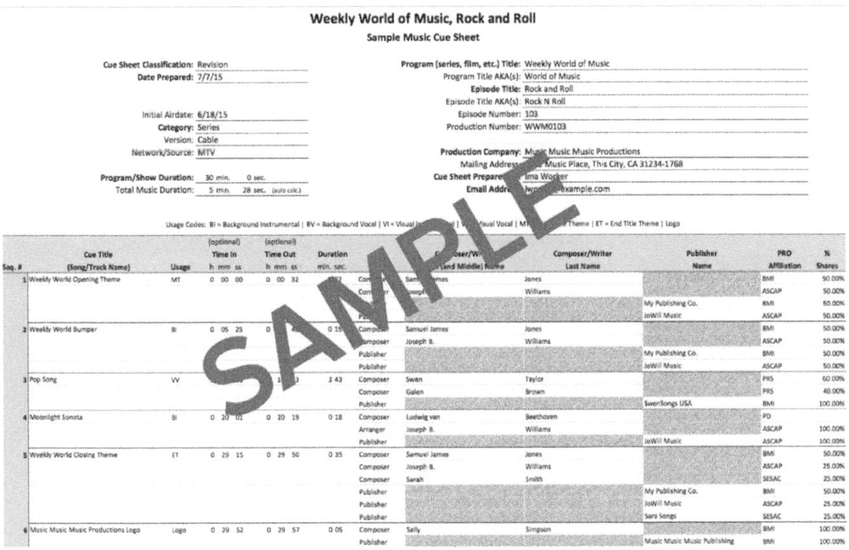

Weekly World of Music, Rock and Roll

Sample Music Cue Sheet

Cue Sheet Classification: Revision
Date Prepared: 7/7/15

Initial Airdate: 6/18/15
Category: Series
Version: Cable
Network/Source: MTV

Program/Show Duration: 30 min. 0 sec.
Total Music Duration: 5 min. 28 sec. (auto calc.)

Program (series, film, etc.) Title: Weekly World of Music
Program Title AKA(s): World of Music
Episode Title: Rock and Roll
Episode Title AKA(s): Rock N Roll
Episode Number: 103
Production Number: WWM0103

Production Company: Mu[illegible] Music Music Productions
Mailing Addres[illegible] Music Place, This City, CA 31234-1768
Cue Sheet Prepare[illegible] Ima Wo[illegible]er
Email Addr[illegible] Iw[illegible]example.com

Usage Codes: BI = Background Instrumental | BV = Background Vocal | VI = Visual [illegible] | [illegible]sual Vocal | M[illegible] Theme | ET = End Title Theme | Logo

Seq. #	Cue Title (Song/Track Name)	Usage	(optional) Time In h mm ss	(optional) Time Out h mm ss	Duration min. sec.		Composer/Writer First (and Middle) Name	Composer/Writer Last Name	Publisher Name	PRO Affiliation	% Shares
1	Weekly World Opening Theme	MT	0 00 00	0 00 32	[illegible]	Com[illegible]	Sam[illegible]mes	Jones		BMI	50.00%
						Com[illegible]r	[illegible]oseph	Williams		ASCAP	50.00%
						[illegible]			My Publishing Co.	BMI	50.00%
						Pu[illegible]			JoWill Music	ASCAP	50.00%
2	Weekly World Bumper	BI	0 05 25	0 [illegible]	0 15	Compo[illegible]	Samuel James	Jones		BMI	50.00%
						[illegible]mposer	Joseph B.	Williams		ASCAP	50.00%
						Publisher			My Publishing Co.	BMI	50.00%
						Publisher			JoWill Music	ASCAP	50.00%
3	Pop Song	VV	[illegible]	[illegible] 1[illegible]3	3 43	Composer	Swen	Taylor		PRS	60.00%
						Composer	Galen	Brown		PRS	40.00%
						Publisher			SwenSongs USA	BMI	100.00%
4	Moonlight Sonata	BI	0 20 01	0 20 19	0 18	Composer	Ludwig van	Beethoven		PD	
						Arranger	Joseph B.	Williams		ASCAP	100.00%
						Publisher			JoWill Music	ASCAP	100.00%
5	Weekly World Closing Theme	ET	0 29 15	0 29 50	0 35	Composer	Samuel James	Jones		BMI	50.00%
						Composer	Joseph B.	Williams		ASCAP	25.00%
						Composer	Sarah	Smith		SESAC	25.00%
						Publisher			My Publishing Co.	BMI	50.00%
						Publisher			JoWill Music	ASCAP	25.00%
						Publisher			Sara Songs	SESAC	25.00%
6	Music Music Music Productions Logo	Logo	0 29 52	0 29 57	0 05	Composer	Sally	Simpson		BMI	100.00%
						Publisher			Music Music Music Publishing	BMI	100.00%

SAMPLE

STUFF TO REMEMBER

- Music is perhaps the most powerful way of communicating emotion in storytelling.
- Temp music is used in the editorial process to establish pacing of the edit.
- Rhythmic music can be edited seamlessly if edits are made on the beat.
- Every piece of music used in movies or television must be listed on a cue sheet.

PUTTING IT INTO PRACTICE

- Try searching the Internet to find production music that conveys a specific emotion.
- Edit a favorite piece of music to make it half as long as the original in a way where no one will notice where the edit happened.

11 REVERBS AND DELAYS

GIVE IT SOME SPACE

Reverberation (or "reverb") is the continuation or reflection of sound after it is produced. It is the sound of a large room, a cave, or a church. It's the sound of the room. Major orchestras invest a large amount of resources to create a great sounding hall in which they can perform and where—as is the case also in theaters—sound is projected out to the audience.

Sound waves bounce off of hard surfaces like light bounces off a mirror, or a cue ball bounces off the cushion on a billiard table. The bounce is spherical, scattering at every reflection, building upon itself, and creating a complex "ball" of reflections that we hear as reverb. What occurs in a live-sounding reverberant room is multiple reflections building upon one another.

Reverb is also frequency dependent—high frequencies are more directional then low frequencies—and is affected too by absorptive properties of the surfaces in the room. A church will have a darker sound than a skyscraper lobby. Mentioning these two particular rooms might give you an idea of why we are spending a chapter talking about reverbs and delays. In audio post-production, we use reverb to compensate for the lack of space that existed during the shoot of the film. If we want to audience to believe that they are in an imaginary space, we have to create the space.

BUT FIRST, HOW WE GOT HERE

A Real Space

Reverbs are used extensively in music production and the first reverbs created used physical spaces. You can create a room reverb by placing a speaker at one end of a room, playing the audio that you want to add reverb, and a microphone at the other end of the room in order record it. A famous example of this is the Capitol Records building in Los Angeles. During the construction of the building in the mid-1950s, cavernous rooms—"the lungs of Capitol"—were built thirty feet below the surface and it was there that many famous recordings were created in the studios of the building.

Plate reverb

Eventually, an artificial reverb was created, the plate reverb. A metal plate was suspended in a closed box and was "excited" by placing a speaker transducer in the center of the plate. The characteristic vibration of the plate was picked up by microphones placed on the corners, and became an excellent reverb for vocals.

Spring

A spring reverb is a smaller, cheaper version of a plate reverb, which uses (as you might guess) a metal spring instead of a metal plate. You can find spring reverbs in guitar amplifiers. In guitar amplifiers, if the reverb is on and you knock on the box of the amp, you can hear the springy reverb vibrating. Spring reverb is a characteristic of the 1960s "surf" sound.

Digital

Eventually, digital versions of plate reverbs were created, even as early as 1976. In the 1990s, more cost-effective versions of digital reverbs were created and today, every possible reverb algorithm has been reproduced and packaged in little metal boxes. You can find reverb sounds characteristic of a hall, church, room, spring, etc.

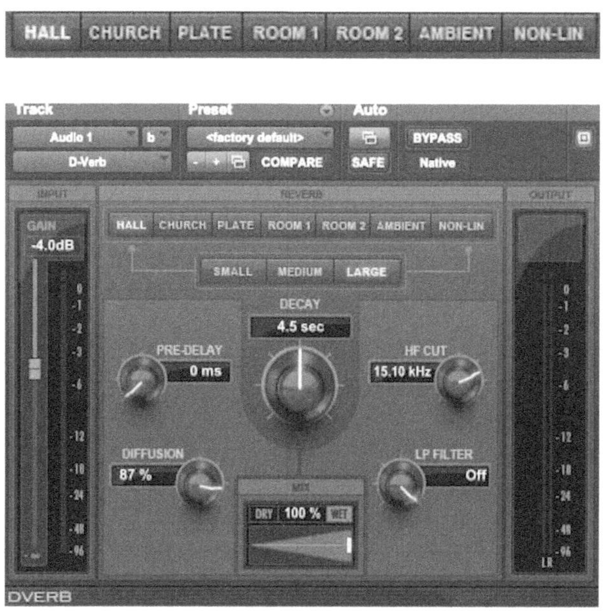

Types of reverb choices in Avid Pro Tools

Reverb choices for plugins by AIR and Waves

Convolution

Convolution reverb is a mathematical recreation of a real environment based on an impulse response (IR). An impulse response can be created by recording a sweep tone (a frequency tone, sweeping from the lowest to the highest), a starter gun, (which is a burst of all frequencies), or even the "crack" of a snare drum. There are programs that will convert this recording into a reproduction of the space. The companies that create convolution reverbs also come with a wealth of useful convolution spaces usable in audio post. Interior spaces included are domestic rooms like living rooms, bedrooms, or kitchens. Exterior spaces usually include caves, forests, or parking lots. Consider the possibilities of recording impulse responses in the same space that the film was shot, as this will enable you to alter ADR recordings and make them sound like the exact space that the film was recorded. When I interact with the production sound crew, I always ask them to record a filter sweep in unusual spaces, since I won't know during production if I'm going to need that space. I would rather have an option just in case I need to solve a problem later on.

Altiverb's Convolution Stairway library

USING REVERBS IN POST-PRODUCTION

To use reverbs in audio post-production, as with other processes you have to tread lightly so that the audience doesn't realize you are creating something that is not real. If it sounds like you are putting reverb on a voice, for example, the audience won't believe it is true to the story, and like everything else that is over-done or under-done, it will take them out of the viewing experience. Whether it is a parking lot or an empty building, the audience needs to believe that they in the space they see on-screen. Also be aware of reverb clichés to avoid, such as: indicating when someone is thinking; a flashback sequence; dreams; and or gunshots.

DELAYS

Delays are a repeat of sound. It is the typical effect that you get when you yell "Hello!" in a canyon and a *"Hello!"* comes back to you a second later. What causes that effect is a single reflection: It is the sound bouncing off of a hard surface and returning to you. This bouncing around of sound is part of our everyday lives, but the repeated sound is not so obvious. That said, if you employ a little critical listening to spaces, you will learn to pick up on these reflections.

Place yourself in the middle of a large room with hard surfaces. Clap your hand once and listen for the reflection coming from a wall. It will be fast, maybe 50 milliseconds, depending on the size of the room.

Now try going outside, in a parking lot that is next to a building. Clap your hands and you can hear the reflection come back to you from the building. That reflection is much slower than what was heard in a room, because the distance the sound had to travel is greater and thus the sound took more time to get back to you. These kind of first reflection reverbs are critical in making environments sound real.

Now let's try some critical listening to *how* the reflection sounds. And as we listen, let's apply what we hear to a delay plugin. First, here is a simple free delay plugin that comes with Avid Pro Tools.

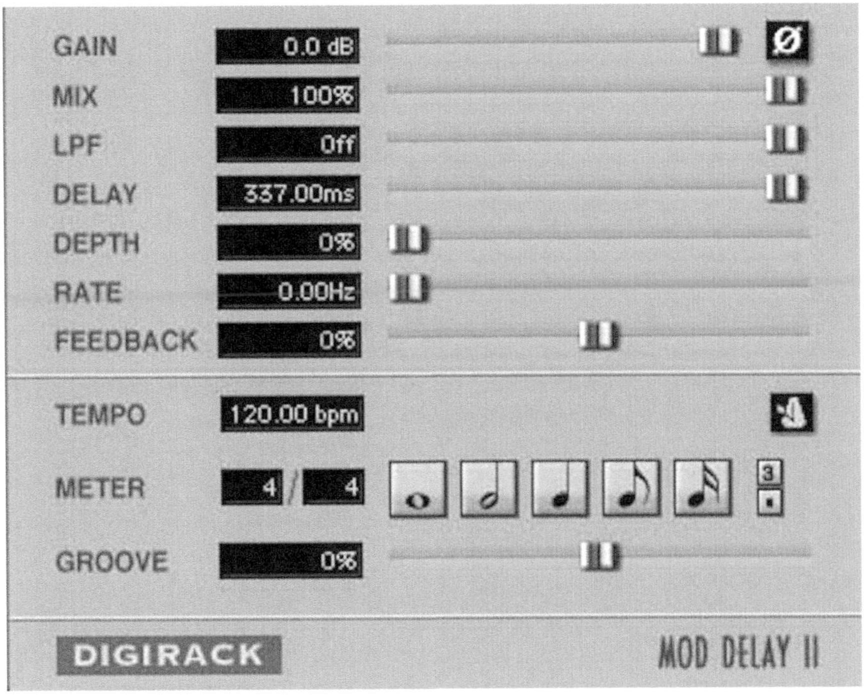

Here are the important basic controls that can be found on just about any delay plugin:

- Gain—the level adjustment as audio passes through the effect;
- Mix—the balance of the input source and delay audio;
- LPF—a low-pass filter to remove high frequencies;
- Delay—the time in milliseconds between the source and the delay;
- Feedback—the delay signal re-delaying, creating multiple repeats.

Let's try and recreate a real-word scenario that has significant delays and reflections. An excellent example is President Obama's first acceptance speech in Grant Park, Chicago, in November 2008. You can use any exterior speech situation that will give you a long enough delay to analyze the sound bouncing back to you.

As you listen to the speech, notice the reflection coming back off of the distant buildings. How long is the delay? Adjust it to about 400 milliseconds. How many repeats do you hear? There are actually three, so adjust the feedback so that you hear three repeats.

Now think about the quality of the repeat sound. Delay plugins will just spit back the same audio. The repeat from the speech does sound very different from the source. It sounds thinner, with significantly reduced low frequencies. The plugin does have an LPF, which will reduce the high frequencies, but this is the opposite of what we want to do, so we need to insert an EQ (equalizer) after the delay plugin, which will affect the delay and give us more options for altering frequencies than just a low-pass filter.

Then you can apply a HPF (high-pass filter), which will allow through the high frequencies and remove the low frequencies.

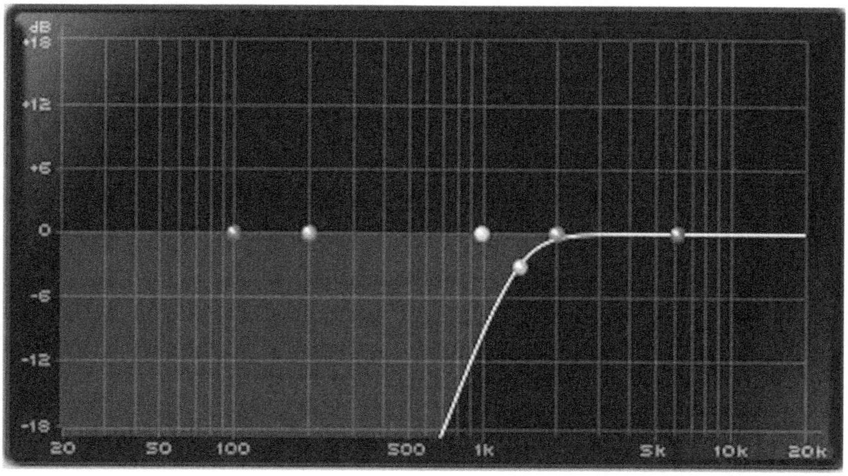

How does it sound now? It's getting pretty close. You will find that these critical listening exercises are useful in reproducing real environments while retaining complete control of your audio. Now that you have gone through this exercise and recreated reality, your perception of delays and reflections will have been improved. However, in filmmaking, reality is not always the choice. Some very successful films that contain scenes very similar to our Obama speech example do not sound anything like reality. You may not want reality and choose to have a little more clarity of dialog instead. Try comparing President Whitmore's speech in *Independence Day* (which sounds real) with the funeral speech in *The Dark Knight* (which does not sound real).

Pre-Delay

A pre-delay is a short reflection that is applied to reverbs in order to delay them fractionally, usually for between one and 300 milliseconds. This gives us a space before we hear the reverb effect and yields a very realistic effect, since many interior spaces have a combination of a first reflection (that we hear as a delay), and the reverberation characteristic of a large room.

AUX CHANNELS—OR LET'S TAKE A BUS

Let's create a scenario where we put three or four tracks of dialog into a reverb space. How would you go about doing this? Your first impulse is to insert identical reverb plugins on each track, but if you do that and decide later to adjust a parameter of the reverb, you'll then have to go back and adjust each individual plugin. You want them all in the *same* space, ideally, and there is a better way to do this which happily also has less impact on the CPU usage of the computer. (Reverbs are CPU hogs.)

Revisiting the "lungs of Capitol" mentioned above, the engineer sent vocals to one room in the basement. And that's just what we want to do here—send the dialog tracks to one room. This way, we are using only one reverb, and when we adjust that, we adjust the effect on *all* of the dialog tracks. We are going to use a separate internal path for audio; a separate fader or *bus send,* and an *auxiliary track*.

The idea of a bus send and an auxiliary track is sometimes difficult to wrap your head around. Auxiliary tracks (aux tracks) are the place where audio is sent to be affected by audio processors (plug-ins) like reverbs, delays, etc. In our example, the aux track is the basement room of Capitol. The bus is the cable going down to the basement. The bus is the path, and the aux is the destination. From there, the aux track will go on somewhere else, in our example, back to the control room of the studio and out to the main speakers.

Let's put some space on our dialog tracks. First we create an aux track and insert a reverb plugin. Its input (going to the aux) is Bus 1.

Now on each of the dialog tracks, we choose Bus 1 in the Sends.

We see that a small fader shows up on the channel.

This fader adjusts how much level we are going to send to the aux, or how much reverb do we want to apply to that individual track.

The fader on the aux track will adjust the overall reverb.

Now if you want to make a change (small or large) to the reverb, you change a setting on the reverb plugin inserted in the aux track, and you will affect the space of all of the dialog tracks. You could even change plugins, or add in a series of effects for a change as an experiment, and you still have to deal with only one track that processes them all.

If you have ever played in a band or done PA work for one, you'll know that the PA system has a rotary pot on each channel for monitors. If you turn the monitor send up, that microphone gets more level in the monitor. And when you adjust the level of each microphone, you are making a separate mix to the stage monitor. That monitor mix is an aux track and the monitor send on the mixing board is a bus.

Auxiliary tracks can be invaluable in the mixing process and once you familiarize yourself with a quick setup, you will never go back.

STUFF TO REMEMBER

- Reverb plugins are used to simulate real spaces.
- Reverbs can be used in creative ways, although there are clichés to avoid.
- Delays are as important for creating space as reverbs.
- Auxiliary tracks are the best way to use reverbs and delays in an efficient and consistent manner.

PUTTING IT INTO PRACTICE

- Try recreating an exterior speech using reverbs and delays, much like we did in this chapter.

---

12 MIXING
Putting It All Together

Now we have come to the time that we've been waiting for. Well, at least it's what I wait for. And what I'm waiting for is sitting in front of me. After weeks or months of collecting sound effect, editing slivers of audio, and cleaning up dialog, you can put it all together. In fact so much time may have passed that you may not remember why you made some of the thousands of decisions taken before you sat down in front of it. "Why did I choose the lavalier mic over the shotgun?" "What is that extra layer on the sound effects?" "I know the composer is competent, but his score is taking up so much sonic space, there's no room for anything else." That's right, it's time to mix!

Mixing involves combining all of the audio elements of the movie and balancing the levels in such a way that the audience does not notice that you have created an artificial world. Everything should sound like it all belongs together in one cohesive audio framework that transforms the audience to another place. This may seem like a pretty daunting task since you may have 100 tracks in front of you. Usually on feature films, my track count is roughly 100, and I try to keep it around that number. In Hollywood studio films, there could be *hundreds* of tracks; so many that it takes more than one person to mix the film. Three rerecording mixes often happen on projects of that scale: One each for dialog, effects, and music. Usually the dialog mixer is the rerecording mixer, the person responsible for the final delivered mix. The sound effects mixer is usually the supervising sound editor, the person responsible for overseeing and managing all sound effects recordings.

On independent films, however, it is usually a single individual who will take on all of these tasks solo. As I mentioned earlier in this text, I am also a composer and sound designer. This means that, other than production sound recording, I have created every piece of audio in the film. I find the process personally rewarding. I also think that it is an efficient way to work as it affords a singular vision of the sonic characteristics of the film. Directors seem to find the process of working with one person streamlined and productive. They have to communicate with just one individual and the working relationship strengthens over time. Never underestimate the relationship with the director. (In the case of television, that person in charge is usually the producer.)

So how do we approach mixing all of these elements? Organization, what else? For each stage of audio post, I usually keep separate sessions; one each for dialog, foley, footsteps, hard sound effects, ambiences, and music. When it is time for the final mix, I can bring all of these separate sessions into a single master mixing session. This way, if I ever make an error, which I often do, I can go back to one of the element sessions and retrieve what I need.

And, as I bring in each element session into the master, I keep them grouped together. So from top to bottom, there is dialog, ADR, NDP (non dialog production), foley, footsteps, hard sound effects, and music. I pretty much keep to the same layout in all projects, so after a while it becomes second nature to me if I want to find something. I also color my tracks so I can quickly find what I am looking for. As with grouping, I keep the same colors for every type of track from project to project. For example, dialog is always yellow, ambiences are always pink, hard effects are always purple, etc.

Here is what a typical session looks like. This is the master mixing session for the film *Death House*.

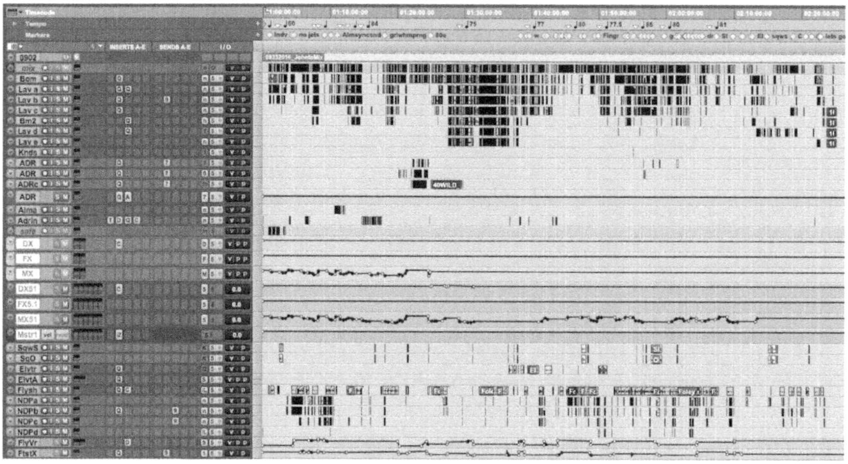

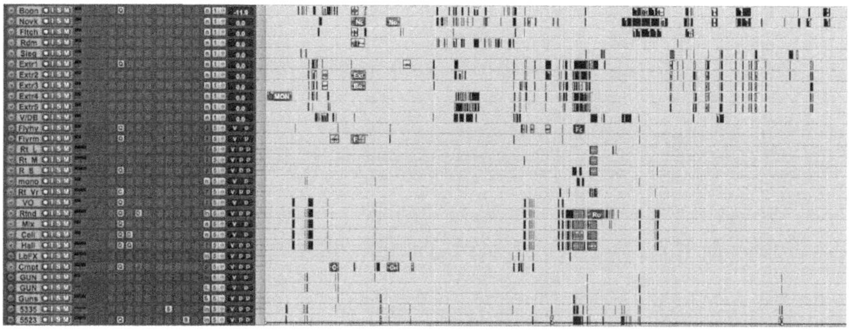

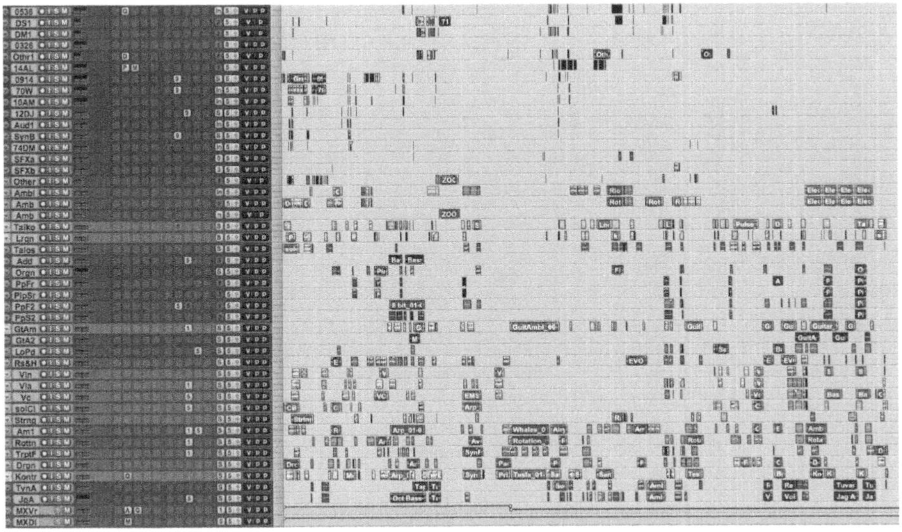

Below the dialog and ADR tracks is the master session, which includes separate faders each for dialog, effects, and music, and then a master fader. Each of these three groups is called a *stem*. Together the three stems make up the final mix, sometimes *called print master*. I created separate faders for dialog, effects, and music (DME) by using auxiliary tracks for the three stems: All of the dialog tracks are bussed to the dialog aux track, and the same happens for effects and music. (Aux tracks and busses were discussed in Chapter 11.)

Sometimes I use aux tracks going to aux tracks. For example, if I have eleven footsteps tracks, rather than bussing them to the effects aux, I can create a footsteps aux where I can filter out low frequencies for all of the footsteps by inserting a high-pass filter. The output of the footsteps aux is then the effects aux.

I go about balancing all of the tracks as I work on one scene at a time, making it a single cohesive unit. Most of the time, all of the elements sit pretty well in the mix and I need to adjust only the three stem faders. These faders are writing my moves as I move the fader up and down. This is called *automation*.

AUTOMATION

Each channel has the possibility of volume automation, which means that the DAW will remember the changes you made so that when you play the film back, you can see the faders move. You can make this happen by writing the volume changes with a mouse as you look at the *volume clothesline* for each track. A "clothesline" is a term for the horizontal line on top of the track that is a representation of the position of the fader. You can change the fader position by pulling on the line in a similar way as you would pull on a clothesline. The figure below depicts clothesline automation.

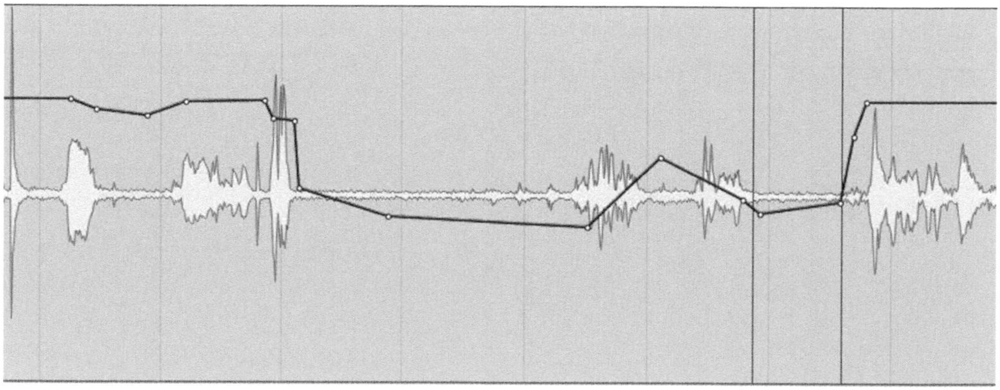

Or you can have the DAW record your moves of the fader when it is in "Write" mode on the automation tab.

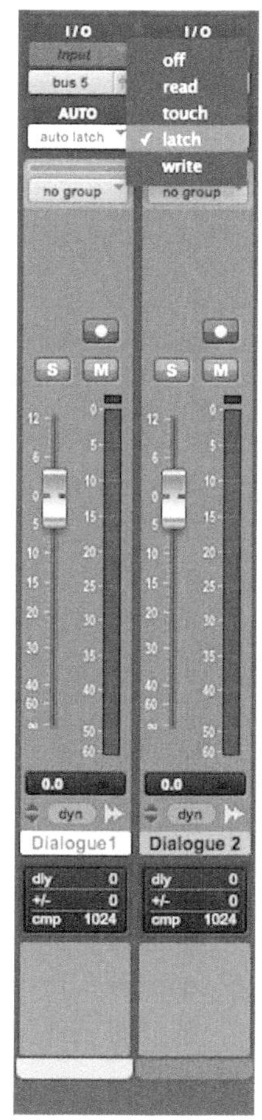

The automation tab gives you several options.

- **Off.** The DAW will not play back automation and the fader will stay where it is.
- **Read.** The DAW plays back the volume changes.
- **Write.** You write, or write over, the changes in the fader movement.
- **Touch.** When you touch the fader with the mouse, it is writing your moves, but when you release the mouse the fader will go back to playing the previous fader movement.
- **Latch.** When you touch the fader with the mouse, it will record your movement, but when you release the fade, the volume with stay ("latch") in place where you left it.

Before I even place one instance of volume automation, I like to balance all of the tracks in the general area of where they sit in the best in relationship to each other. I may play through the entire film once or twice while adjusting the faders to get an overall feel of the whole film. Then I will start for real with the first scene, again mostly using the DME stem faders.

With dialog being king, all other elements are secondary to it. And if we did our dialog editing correctly, the dialog will sit at the correct level and the volume of the other two stems will be balanced to the volume of the dialog. If there is no dialog in a scene, or there is an extended period without dialog, something else is usually being heard. If it a music-driven scene, the effects will be under the music; if it an effects-driven scene, the music will be subordinate to the effects. Whatever the situation, all three stems cannot be equal. One of them has to be the primary.

What I have learned is that having the right amount of time to mix is important. Our ears are muscles, and muscles get tired. I have sat with directors and mixed for fourteen hours straight, probably because of some festival deadline. After that amount of time, nothing sounds good and you cannot trust your ears. What works is multiple short mixing sessions. Usually I will work at my own pace of a few hours a day and get ready for a presentation to the director. We don't necessarily have to go through the whole film and I will set that agenda of informing him of my intentions of the mix. Maybe it will be the first half hour. Then, maybe, if the director and I have a decent relationship, and he/she is able to listen to a rough mix, we will go a little further and I will ask questions. The goal is to schedule a final mix with fresh ears and make tiny adjustments. I have done this in as little as two hours (for a ninety-minute film) in the morning just before the director was leaving for the airport. We listened for the overall level of the dialog, while adjusting a scene up or down a little. That was a luxury.

COPIES

I feel I have to mention backups. When I spend a couple of hours mixing a scene, the last task that I do is copy the entire session to the projects backup drive. Both the work drive and the backup drive are identical. Periodically I may make a third copy on an archive drive that contains my projects for the year. If there's a backup, I can sleep better at night.

WHAT TO CHARGE

Until you've worked through the audio production process for an entire film, it can be pretty hard to guesstimate how long it might take you. (Naturally the more projects you work on, the clearer the idea you'll have.) But take a minute now, at the end of the project and think about what you put into it. It is very possible that you have spent months on a single film. (Hopefully you have done other things at the same time.)

So what should you charge? I believe that there are three reasons for taking on a particular job. One, because you want to be involved with that specific project (because it is so good). Two, because you want build a working relationship with the filmmakers. Three, because you want to make some money. If you happen to stumble on something that satisfies all three considerations at the same time, you are a lucky individual. Most of the time, however, it is about compensation. Should you charge a percentage of the film's total budget? I often ask what the total budget is. That won't work if it is $1,000, though, and I doubt it you would get it if it were $100,000,000. Decide what it is worth to you. In independent film, however, many filmmakers will not know the importance of what you do. It is your responsibility to educate them, by showing them how great you will make their masterpiece—not lecturing them about it. Develop a relationship. But get what you believe that you are worth in compensation.

STUFF TO REMEMBER

- Group your tracks by type and use color coding.
- Avoid mixing for long stretches of time, which creates ear fatigue.
- Use automation only after tracks are balanced roughly where you want them.

---

13 DELIVERABLES

YOU MAY THINK THAT YOU'RE DONE . . .

The mix is done.

The director is happy and thinks that you saved the film.

But a new player is coming on to the scene and they have a lot of power. You will probably not have direct communication with this player, meaning the director will be an anxious middle-man. When you were mixing the film, solving problems, and realizing the director's vision, these people probably hadn't even heard of it. Who am I talking about? The distributor.

Many low-budget independent films seek distribution after the movie is completed. The filmmaker may actively reach out to distributors, or decide to submit to festivals and market their film through those possibilities. The film may get distribution as early as the week after you have completed the mix, or a deal may happen only two years later, when you are in the middle of another important project.

What is a distributor? A distributor is responsible for marketing the film and finding an audience. They may also be responsible for the finance of the project so that the film can get made in the first place. In low-budget movie production, the filmmaker usually gets the film financed on his or her own, and then makes an agreement with a distributor to market and book the film for release, which could be anything from DVD/Blu-ray to video on demand, streaming, or appearing in movie theaters. The distributor, like everyone else, wants to make money while doing as little work as possible on the production of the film.

So how does this relate to audio post-production? Two reasons: the deliverables; and Quality Control.

DELIVERABLES

In any agreement with a distributor, there is a list of deliverables which specifies a wide range of elements, but for our purposes the most important part of that will be any audio requirements. Very often the director will be nervous and want to understand what the distributor is asking for since "it's all Greek to me." It is part of your job, and good for your career, to put the director at ease and explain to them what is needed.

To give you some insight to what deliverables are, below are the audio specs for three features that I have worked on recently.

Feature #1

If you are delivering 5.1 sound, you must include TWO Pro-res Masters of your feature - one with 5.1 embedded as noted above and another one with Stereo embedded. Not all platforms take 5.1

All audio should always include a stereo mix. Files delivered with surround should also contain the 2.0 mix after the 5.1.

Codec	L-PCM
Sample Rate	48kHz
Bits per Sample	16/24 bit
Audio Configuration	**Stereo(2.0)** – 1 Left: 2 Right **Surround(5.1 +2.0)** – 1 Left; 2 Right; 3 Center; 4 LFE; 5 Left Surround; 6 Right Surround; 7 Stereo Left; 8 Stereo Right or 1 Left; 2 Right; 3 Center; 4 LFE; 5 Left Surround; 6 Right Surround; 7 Stereo (interleaved)

Feature #2

b) Sound Elements: (Please refer to Exhibit X-3 for Specs)

i) Firewire Drive which contains the Pro Tools sessions of the Theatrical Mix, which should include 6 track Printmaster, 2 Track Printmaster, 6 Track M&E, 2 Track M&E, DME, all 5.1 stems (Dialogue, Music and Effects). (Compressed Files are unacceptable)

ii) Television Version:

A) One DVD-Rr 2-track printmaster and one DVD-R 6-track printmaster of the television version (Left Total, Right Total & Left, Right, Center, Subwoofer, Left Surround, Right Surround), which integrates all loop lines for TV language censorship purposes.

B) One DVD-R 2-track and 6-track fully filled in Television version M&E (Left Total, Right Total; Left, Right, Center, Subwoofer, Left Surround, Right Surround). See attached Exhibit "X-3" for specifications.

C) An extensive, detailed description of the changes made in the theatrical feature to create the television version, including, but not limited to, notations as to footages where changes were made, length of changes, and picture descriptions.

D) Closed Caption files plus typewritten copies of all song lyrics contained in the Production.

iii) All printmasters, TV versions and M&E's (compressed digital files not acceptable).

mix stems and all premix units on a firewire drive in Pro Tools format used to create the

iv) Music:

A) One (1) CD of all music cues (source and score) as used in the final Production assembled in order as they appear in the Production with a list and timings.

B) One (1) CD of all score music.

C) One (1) CD of all source music (source song in its entirety, unedited.)

Feature #3

2. HD Mezzanine File

- **Apple ProRes 422 (HQ)**
- **ITU-R BT.709 color space, file tagged correctly as 709**
- **VBR expected at 88-220 Mbps**
- **1920 x 1080 square pixel aspect ratio material***
- **23.976, 24, 25, 29.97 frame rates are supported**
- **Native frame rate of original source:**
  - o **Fields and frames may not be duplicated or eliminated to create a broadcast frame rate (for example, telecine, NTSC to PAL conversion)**

AUDIO:

Surround

- **LPCM in either Big Endian or Little Endian, 16-bit or 24-bit, at least 48kHz**

1- L
2- R
3- C
4- LFE
5- Ls
6- Rs
7- Stereo L
8- Stereo R

- **All tracks should be PCM Little Endian**
- **Separate 5.1 audio is accepted as .WAV files sent as individual tracks.**

Stereo

- **LPCM in either Big Endian or Little Endian, 16-bit or 24-bit, at least 48kHz**
- **Expected Dolby Pro Logic channels: Lt, Rt or expected stereo channels: L, R**

What similarities do you find between the three examples? I see that they all want Linear PCM WAVs, which is what all of our DAWs output. Probably what they are saying is that they do not want MP3s, which we wouldn't dare deliver since the MP3 compression changes the sound. While very popular because they take up one-tenth the space as WAV files, MP3 do not have as high a resolution. One of the many qualities lost due to the compression is that MP3s only reproduce frequencies up to 18 kHz while WAV files reproduce frequencies up to 24 kHz.

Sample rate is 48 kHz, as expected, but they do not have a specific preference for bit depth, requiring 16 or 24 bit.

When delivering surround channels, they require individual mono tracks that will make up the six channels in 5.1. Placement order is similar, and when the editor or colorist is making the deliverable video file, they should be in the following order of L, R, C, LFE, Ls, Rs, and stereo.

Some dated items are listed: I believe some distributors have been copying and pasting the same document for years, and that might explain why they are asking for a firewire drive, even though firewire connection on computers started disappearing in late 2012. Music delivery on CDs is also dated, but I understand why they would want unmixed source music, and unmixed music. Remember the distributor is involved in marketing the film and they may want to create their own trailer, and so need all music tracks in their entirety.

If you have planned your mix correctly and saved copies, you should be able to access what is needed in short order, making your director very happy.

QUALITY CONTROL

So all of the necessary files have been sent to the distributor and everyone is a little more at ease as a result. Then you get another call from the director saying a quality control (QC) report has been delivered and there are sixty-four audio issues on the sheet that have to be resolved by the end of the week or the film will miss the distributor's deadline.

I was a bit taken back when I got my first QC report. I felt that I hadn't been successful at my job even though I know I'd done my best. I felt like I couldn't even trust myself. However, after speaking to a friend who works at a post house that does a lot of color correcting for features, I felt much better. He told me that every film generates a QC list like mine. I was taken aback—how is that possible? What happened to getting it right first time?—but I tackled the list, and it wasn't that bad after all.

Let's look at a portion of a quality control list and decipher what is happening.

In the first column is the timecode of the offending lack of quality. The next column has a V for visual or an A for audio, my area of interest. There is then a description of the problem followed a couple of empty columns later by a 1, 2, 3, or an FYI (for your information). These indications tell us how severe the infraction is, with 1 being the smallest and 3 the worst. All 3s have to be fixed, with 1s almost considered an FYI. You'll also notice some 2+ ratings. I fix those also.

It is also in the best interest of the project and your continuing relationship with the director to resolve these issues on your first attempt—the last thing you want is another QC report to come back highlighting the same problems. Each report will cost the production $1,200 (cost in 2013) that has to be recouped before any monies are to be paid by the distributor.

01:14:34:03	V	BLACKS ARE MORE ELEVATED THAN PRIOR USE OF ANGLE			[FYI]		
01:14:34:14	A	MAN IS MOVING MOUTH BUT NO DIALOGUE IS HEARD			[FYI]		
01:14:56:08	A	STEREO BECOMES DUAL MONO / AUDIO ONLY IN CENTER CHANNEL THROUGH 01:16:37:06			2+		
01:15:51:20	V	WOMAN'S HEAD SEEN BOTTOM OF FRAME/MOSTLY MATTED OUT IN 2.35 VERSION		178	1	THIS ANGLE IN SEQUENCE - LIKELY OK TO FYI	
01:16:29:00	A	LIGHTER EFFECT IS 1 FRAME LATE IN 1.78:1 & 2 FRAMES LATE IN 2.35:1 PER 1 FRAME OFFSET			2		
01:16:59:19	A	GUN SHOT IS 2 FRAMES LATE GOING BY TRIGGER PULL			2		
01:16:59:20	V	COMPOSITE ERROR			[FYI]	NO ENTRY WOUND OR MUZZLE FLASH FOR GUN SHOT	
01:17:09:09	A	AUDIO BUMP ON "TO THIS"			[FYI]	POSSIBLY AS SPOKEN	
01:17:17:10	A	AUDIO TICK/MORE APPARENT IN M&E THAN COMP			[FYI]		
01:17:44:06	A	WOMAN'S DIALOGUE SYNC LOOKS IMPROVED/1 FRAME LATE ON 2.35:1 PER OFFSET			[FYI]	"I GUESS WE SHOULD GO INSPECT GATE THREE, THEN"	
01:17:59:11	A	EFFECT OF ATV NOT COVERED IN M&E			2+	EXAMPLE OF M&E NOT FULLY FILLED	
01:17:59:22	A	OFF-CAMERA LOW LEVEL DIALOGUE "ARE YOU FRIENDS WITH TONY..."			[FYI]	LIKELY PASSENGER IN ATV - CONFIRM INTENT AS PLACEMENT SEEMS ODD	
01:18:56:14	A	AUDIO TICK			[FYI]		
01:19:02:21	V	BOOM IS VISIBLE IN TOP OF FRAME BEHIND FENCE		178	2+		
01:19:14:00	V	BOOM IS VISIBLE IN TOP OF FRAME BEHIND FENCE/ABOVE MAN ON RIGHT		178	2+		
01:19:18:00	V	MAN ON RIGHT THAT IS TALKING IS CROPPED BY RIGHT EDGE OF FRAME		178	2		
01:19:28:21	V	BOOM IS VISIBLE IN TOP OF FRAME BEHIND FENCE/ABOVE MAN ON RIGHT		178	2+	INTERMITTENTLY DIPS INTO FRAME THROUGH SHOT	
01:19:31:05	A	AUDIO TICK IN COMP AND M&E			[FYI]		
01:21:03:04	A	HAND TO CHEST EFFECT IS 2 FRAMES LATE			2	AS BEFORE BUT 3 FRAMES LATE IN 2.35:1 VERSION PER OFFSET	
01:22:02:23	A	AUDIO TICK OR LOUD TONGUE SMACK			[FYI]		
01:22:13:09	V	SHOT IS MORE WASHED OUT THAN PRIOR USES OF ANGLE			[FYI]		
01:22:32:07	A	AUDIO TICK			[FYI]		
01:23:19:05	V	SLIGHT ENCROACHMENT BOTTOM MATTE FOR 7 FRAMES		178	2+		
01:23:22:04	V	SLIGHT ENCROACHMENT BOTTOM MATTE FOR 6 FRAMES		178	2+		
01:23:32:19	V	ENCROACHMENT LOWER-LEFT AND UPPER-RIGHT CORNERS		178	2+	01:24:09:15 – ENCROACHMENT LOWER-LEFT AND UPPER-RIGHT CORNERS/PER Z-AXIS REPOSITION	
01:23:57:04	V	SLIGHT ENCROACHMENT BOTTOM MATTE FOR 6 FRAMES		178	2+		
01:24:09:15	V	ENCROACHMENT LOWER-LEFT AND UPPER-RIGHT CORNERS		178	2+	PER Z-AXIS PICTURE REPOSITION/MATTED OUT IN LETTERBOX VERSION	
01:24:25:15	V	ENCROACHMENT LOWER-LEFT AND UPPER-RIGHT CORNERS		178	2+	PER Z-AXIS PICTURE REPOSITION/MATTED OUT IN LETTERBOX VERSION	
01:24:55:21	A	AUDIO TICK			[FYI]		
01:25:10:16	V	ENCROACHMENT LOWER-LEFT AND UPPER-RIGHT CORNERS		178	2+	PER Z-AXIS PICTURE REPOSITION/MATTED OUT IN LETTERBOX VERSION	
01:25:16:17	V	ENCROACHMENT LOWER-LEFT AND UPPER-RIGHT CORNERS		178	2+	PER Z-AXIS PICTURE REPOSITION/MATTED OUT IN LETTERBOX VERSION	
	V	ENCROACHMENT LOWER-LEFT AND UPPER-		178	2+	PER Z-AXIS PICTURE REPOSITION/MATTED OUT	

Let's look at a few examples in more detail, and forgive me if I grumble a little at times.

1:14:56:08 *2+* *Stereo becomes dual mono/Audio only in Center channel*

This is the biggest issue to resolve, as it has a 2+ rating. What they are saying is that in 5.1 surround, there is only audio in the center channel, and when I create a stereo version, that center channel is now identical in the left and right—dual mono. I know the scene that they are talking about and I know how the production sound crew recorded it. It is a scene of a mock gunfight using paintball guns, and the audio was recorded with one shotgun mic on a boompole. That one microphone picked up the gun sounds, the footsteps, the dialog chatters, and the bugs in the woods. Of course it would be one channel since it is one mic. And if it is converted to stereo, it makes sense that it would be dual mono. Yet I have to fix it. What I did was create a track of stereo forest ambience with insects and wind in trees. This solved the problem. In 5.1 surround, there is LCR (left, center, right) audio, and the stereo is truly stereo.

1:16:59:19 *2* *Gun Shot is 2 frames late going by trigger pull*

I usually place gunshots in the film late in the process because muzzle blasts are a visual effect and VFX tend to come in around final mixes. Therefore, I place my gunshots on the frame where the muzzle blasts have been layered in. The QC says that I am two frames late (going by the trigger pull) and not by the muzzle blast. What should I do? It has a 2 rating. What I did was document the situation and I left the sound effect where it was. This action let them know that I considered their comment seriously and didn't just ignore them.

1:21:03:04 *2* *Hand to Chest effect is 2 frames late*

In this scene, two guys are having a conversation when one slaps the other on the chest. QC says it is two frames late. With all of the thousands of decision made in the film, I cannot be sure if I foleyed the slap, or if it was part of the production sound. It could be both. But if it *is* production sound, the slap can't be out of sync if the dialog is in sync. And, if I recorded the slap as an effect, I would have surely placed it where the production sound slap was located. So what did I do? I moved it two frames and smoothed out the edit. There's no point kicking up a fuss about the issue: Get it done and move on.

1:24:55:21 *[FYI]* *audio tick*

There are a lot of these comments about audio ticks. This film was shot in a rural Pennsylvania forest, with characters running through the woods. How would one know if the tick is an audio edit, or a foot stepping on a twig, or an insect? I fix what I can find, make a note of what I cannot find, and move on.

FULLY FILLED-IN MUSIC AND EFFECTS (M&E) TRACKS

Going through a few QC reports has made me realize that the trend in music and effects deliverables is changing. With every film, distributors expect identical, fully filled-in M&E tracks. When I watch relatively recent foreign-language films that have been dubbed into English, I am aware the M&E tracks, especially the foley tracks, are not complete. It often sounds like something is missing. Now, possibly because foreign markets are lucrative, everything must be there: All the sound, just as it appeared in the English-language mix.

While QC reports may seem bad report cards, I find that I gain an understanding of other parties' expectations, and if I act on the reports, they make me better at my craft.

STUFF TO REMEMBER

- Before you start your mix, find out precisely what the deliverables are.
- A film must pass Quality Control before it is accepted by the distributor.
- Not all items on the QC report have to be addressed.

14 CASE STUDIES
A Couple of Real-world Scenarios

In this final chapter, I am going to walk through a couple of films to give you an idea of practical applications of the techniques discussed in this text. I've chosen two completely different types of films. The first, *My Dog Tulip*, is a 2D animated feature based on the 1956 memoir of the same name about a man who rescued a German Shepherd and how the two created a special bond. The book was written by BBC editor, novelist, and memoirist, J. R. Ackerley.

The second film, *Death House*, is billed as "the Expendables of Horror" and has an ensemble cast of horror icons. The original script was written by Gunnar Hansen, who played the central role of Leatherface in *The Texas Chainsaw Massacre*. I have worked with the director, B. Harrison Smith, on five other features.

MY DOG TULIP

As with most animated films, *My Dog Tulip* began with the recording of the voices. The film did not have many voices and about 95 percent of the film is one person telling the story, which was a lot like reading the book. I had worked with the animator, Paul Fierlinger, for a dozen years and most of the projects that we worked on were home produced, meaning that he animated, his wife Sandra did the coloring, and I produced all of the audio. A lot of the voices that we recorded, if they weren't one of us three, were people that we found. Since *Tulip* had a producer, a budget, and actors with names, this movie was not going to be home produced.

DIALOG

The lead voice was Academy Award-winner Christopher Plummer. Although Paul is based in the Philadelphia area Plummer lives in Connecticut, and so we accommodated him and recorded in a New York recording studio.

Recording voice-over for *My Dog Tulip*

I was rather relieved that since the technical responsibility lay with the engineer, I could focus on keeping track of where all of the lines were. We had one day to do this, without pickups or ADR, even though there is usually little ADR in animation. My recommendation was that we record with two microphones; one would be the standard one foot away from the actor's mouth, and the other, a safety, about two and a half feet away. In the event that Plummer needed to shout a line, while the close mic might clip, the distant mic would capture it properly and give a sense of space. Ackerley (Plummer's character) would be yelling at his dog often in the script.

Plummer's recording went without a problem, but we knew that we had to address another issue. In the print edition of *My Dog Tulip*, Ackerley often drew pictures of and wrote poems about his dog, and Paul wanted to have Plummer sing one of these songs. We didn't know if he could sing. (I don't even remember if he sang in *The Sound of Music*). Yet I had to prepare so we could get something recorded. I had a little melody ready, which was printed on paper and recorded so that he could hear it. At the end of the script read, I explained what we needed and played him the demo. "Never mind," he said. "I'll just sing something," and he sang a little tune, which was exactly like the one I wrote, and he had just listened to. Problem solved.

The next step was to take the recording home, sit down with Paul, and edit the read into a linear timeline so that he could animate. The three other actors who had smaller parts were cut in and the levels were smoothed out. Easy stuff. Paul then took the read home to start drawing. Little else was needed until we heard from the producer that Isabella Rossellini wanted to be in the film, since she is a dog lover. There was a small role for a veterinarian in the story, but since no lines were written, Paul created a short dialog from Plummer's read for Ms. Rossellini to record. The question was where to record the new lines. Because of travel plans, Ms. Rossellini's schedule was limited in New York, and we all wanted to accommodate her, so I offered to record the pickup lines in her Manhattan apartment. It was a simple process recorded on a portable laptop setup running Pro Tools and two microphones: A dynamic and a condenser. I used two microphones because while condenser microphones may have better frequency response, and would capture all the sibilance and clarity that I wanted, dynamic microphones sensitivity drops off around two or three feet. If the room that we were recording was very reverberant, though, the dynamic would be a good option. The short recording session took place in her small kitchen with Ms. Rossellini seated at the table and Paul directing her read. I recorded both microphones at the same time so that I could choose the most suitable option when I got home.

Of course, other vocal sounds were needed for the film. Dog vocals, in fact. So while Paul and Sandra were animating, I went off to record dog sounds.

In the story, Tulip was the kind of dog who was constantly barking, so barks were a primary element in the soundtrack. How does one record a clean dog bark? And with Tulip being a German Shepherd, I needed a German Shepherd. I would take any Shepherd I could find: The movers for my house had a guard dog, a client had an aging female, and I would stop people on the street, who would then look at me weirdly. I learned a lot in the process and discovered two important things about dogs: First, they can't act; and second, once you've recorded about three or four barks, you have their entire range. Most of Tulip's vocalizations came from Angel, my mover's guard dog, who was in fact an angel. A simple knock at the door would get her barking. No dog was going to sit still for a stranger and the distance from the "talent" and my mic would be constantly changing. I had to record with two microphones at the same time mounted on a piston grip. The mic for closer recordings was a Cardiod Condensor with a -24 dB Pad on it (Shure). At close range, the sound pressure level of a dog is pretty great and I needed a mic that could handle it. The second mic was used when I was around six to ten feet away, and a shotgun (Sennheiser 416) did the job. I had all of the German Shepherd barks that I would need for the film. Since I required additional dog vocalizations for backgrounds and "speaking" parts, I also recorded other dogs that I found during Paul and Sandra's daily walks with their own three dogs.

The author recording dog vocals for *My Dog Tulip*

THE USE OF EFFECTS IN CREATING THE SOUNDTRACK

Much like a book, a film consists of distinct chapters and the process of building a soundtrack has to be done chapter by chapter, with each one having a separate Pro Tools session. While *My Dog Tulip* was always being driven overall by Plummer reading Ackerley's book, each chapter has a distinct feel based on the characters introduced in it, but music would be the driving force of the *tone* of each chapter, and I'll discuss this further below. When it comes to the balancing of the three elements of a soundtrack—dialog, music, and effects—whichever is number three in priority is usually a distant third, and in the case of *My Dog Tulip*, effects were bottom of the pile. Dialog moved the story, along with Paul's animation, which itself may be telling a secondary story. Then music would set the tone of the chapter. Effects choices were reduced to hard sounds essential to telling the story, such as dog barks, footsteps (only when close in proximity), door slams, machines (trains, motorcycles, etc), and canine bodily fluids. Dog barks came from the recordings of German Shepherds that I had gathered, while footsteps, clothing, and other close-proximity hard effects were easily recorded in the foley process. Anything that was a machine had to come from a library: It is impossible to fake a steam engine or the hissing sound of a vintage kerosene lamp. But when it came to canine bodily fluids, it was time to be creative. *My Dog Tulip* is not the typical Hollywood children's film, and many believe that Ackerley wrote it to shock the British middle class. It is a realistic depiction of owning a dog, including sex, defecation, and a dog giving birth. For most of these sounds, I used a combination of cooked elbow macaroni mixed with pumpkin pie filling. I would create the sounds by squishing the mixture with my fingers, and once I even extruded the concoction through my mouth. Fun stuff.

This film was not a heavily layered soundtrack of the type that you might typically hear in a 3D animated film produced by Pixar. As noted above, *Tulip* was a squiggly line 2D animated film where the prose of the original book was the primary driving force of the narrative. The sounds that were heard were just the primary hard effects essential to the story. There was no need to foley every dog footstep, or add layers of ambiences to create a believable reality.

MUSIC

When Paul Fierlinger animates, he prefers to draw his animations to music, so my next task would be to compose music for a scene before he started to draw. This is a different way of scoring a film than I'm used to, but we worked together well and he would sit at my house and together we would work on ideas. What was helpful was that we had the narration and dialog from the voice recordings as a guide for timing, so it was almost like scoring a radio show. Paul knew what kind of feeling was intended for each scene. The score was driven by melody that was a combination of light jazz and classical chamber music. At least, that was what we thought in the beginning. I think that it is important to note that the film took three and a half years to complete. It may seem like a lot of time, but we are talking about just three people creating an animated feature.

With this amount of development time, the score evolved to include many genres. It seemed that every time we started a new chapter, which often introduced a new character, we tried something new that would identify the character. Captain Pugh would have an historic renaissance score as the story was told around his bucolic farm. The Blandish family music would have a genre of a stuffy British score, heavy with horns. For Ackerley's sister, Nancy, who was possible emotionally unstable, I was employed to play simple clarinet lines. I am not very good on clarinet, but my unrefined performance was a perfect fit to Nancy's character.

I had always intended to replace as many sampled virtual instruments as possible with real musicians, but those recording sessions would not take place until the film was almost complete. I had to prepare sheet music and Pro Tools sessions as we finished chapters because when all was complete, the recording sessions would happen quickly.

The sessions were relatively simple in terms of instrumentation. We started with a brass section, then moved on to some solo strings, before ending with myself on piano accompanying the excellent clarinet work of Arnie Running. One music cue was an oompah marching band piece that was heavy on tuba, and I had the pleasure of recording Brian Brown in my home studio. The cue was used on the end credits.

Below is an excerpt from *The Hudson Review* where Dean Flower explains *Tulip*'s music as a reviewer.

Even more omnipresent than human voices are John Avarese's musical accompaniments—or should I call them commentaries?—in a delightful array of styles. He explains in "Making Tulip" that "Initially, the concept of the film would be string quartet—let's keep this very European-sounding and classical." But as the film developed, so did its improvisatory music. Some of the classical or chamber music elements are still present, but piano jazz interventions are frequent, as are march rhythms with oompah instruments, brief choral or organ interludes, and ¾ time song melodies, plus a stride piano theme inspired by Fats Waller. In this lighthearted, wide-open musical atmosphere I was pleased to hear a reference to "Se vuol ballare" from The Marriage of Figaro. *The moment comes when Ackerley takes Tulip to the country for a visit with his old army friend, Captain Pugh. The expectations of bucolic pleasure are ruined by Tulip's disastrous middle-of-the-night defecation for the occasion. But when I asked Avarese about it, he said the only intention allusion was to the realm of the Renaissance lute—the pavans and galliards of John Dowland. Sure enough, Mozart dissolves into Dowland after a few bars, and the bucolic theme (lutes, shepherds, fair maidens) fits better than Figaro. But I defy anyone who knows the opera not to hear it first as Mozart played on the harpsichord. Either way it's exquisitely sardonic comedy.*

The whole richly-commentative score deserves more discussion, but I will confine myself to a moment early in the film when Ackerley takes Tulip for a walk in a dreary industrial wasteland beside a canal. He has just commented poignantly, as the book does, "It seems to me both touching and strange that she should find the world so wonderful." Indeed we see Tulip delighted by all its rank and rancid smells, intent on sniffing its fascinating messages and leaving some of her own. At this point Plummer sings—or rather whispers, tunelessly—a little ditty that Ackerley wrote but never published. Here are the words, which sneak up on you before you know it:

Piddle, piddle seal and sign,
I'll smell your arse, you smell mine.
Human beings are prudes and bores,
You smell my arse, I'll smell yours.

Immediately a choir of British voices repeats these sacred words, as if they were a requiem mass. The effect is both funny and breathtaking. Although it may seem ironic, it is absolutely serious—a reverent celebration of the world, accepting and praising Tulip's point of view. Avarese recalls capturing the moment this way:

> *Paul and I knew that at the narration recording with Plummer, we needed to get him to song this little poem that Ackerley wrote, so I made up a little melody that he could sing. He really did not want to do it, and when I tried to teach him this little melody, he brushed me aside and said he would just sing something on his own. He did, in fact, sing back almost the same melody that was presented to him. He just sang it wild with no accompaniment. One take, that's all you get. Then I took the recording and arranged around it.*

The repeat was sung by Drexel University's Vocal Jazz Ensemble, using only its female voices (nine of them, faking English accents), and layering the string orchestra accompaniment on the track later. "It took about 10 minutes," Avarese said. "Everything was done in small pieces and put together at my home."

That small-scale simplicity and technological savvy help enormously to sustain the film's lightness and exuberance, even when the story may seem misanthropic.

The Hudson Review, Excerpt from "The Beastliness on *My Dog Tulip*" by Dean Flower. Volume LXV, Number 1, Spring 2012. Copyright © 2012 by The Hudson Review, Inc. ISSN 0018-705X.

THE MIX

Since the film was assembled in chapters, mixing was a simple process of exporting just the music stems of each chapter session into a single master session. The continuous voice dialog tracks were always kept on the master session and not imported from the chapter sessions, ensuring a smooth level of dialog. Had I equalized or compressed any of the voices, the change would have been global. The voices were,

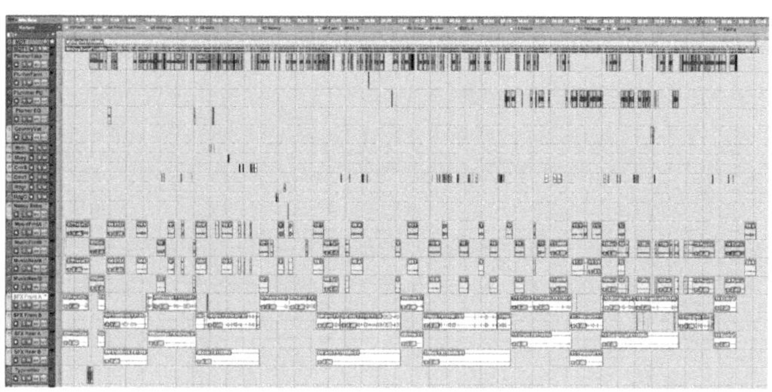

***My Dog Tulip*'s compact master session**

after all, recorded in one day at an excellent voice-over studio, so the levels were consistent throughout. The effects were imported tracks from the chapter session. They were not export mixes but rather all of the individual effect on the original tracks. Once imported to the master session, I could get the track count down to manageable number. Here's a screenshot of the efficient master session.

A STORY ABOUT SURROUND

As the film's production was coming to a close, Paul and the producers thought that it would be a good idea to screen the film. The producers wanted to get feedback from the audience about the story, Paul wanted to see how the film would look projected on a large screen, and I wanted to hear the mix in a theater setting. I had been mixing with near field speakers only four or five feet away from my ears and I wanted to hear the mix with speakers forty to fifty feet away. The screening was held at Dolby's Theater in Manhattan.

The My Dog Tulip Company LLC

You are cordially invited to a screening of *My Dog Tulip* on Tuesday, June 17th at 11:00am.
The screening will be held at The Dolby Screening Room,
1350 Avenue of the Americas (SE corner of 55th Street and Avenue of the Americas), ground floor.

My Dog Tulip is a full length animated film to premiere at the Cannes Film Festival in May 2009. The film, based on the novel *My Dog Tulip* by J.R. Ackerley, has been animated and directed by Paul Fierlinger. We hope to see you.

The screening invitation for *My Dog Tulip* at The Dolby Screening Room

I was pretty excited about the venue since I knew the room would sound fantastic. We loaded the film prior to the screening and checked that everything was playing back properly. The projectionist approached me to enquire if I was pleased with the surround mix. "Surround mix?" I asked. "This is in stereo." The projectionist assured me that the mix was indeed in surround. However, I knew I had delivered only a stereo mix. So what was going on?

A brief word here about Dolby encoding. Dolby encoding does an amazing job of taking the six mono channels of a 5.1 surround mix (Left, Center, Right, Left Surround, Right Surround, Low Frequency Effect), and encoding them into a stereo (2 track) file. When the file is played through a Dolby decoder, it will convert the two tracks back into the six mono surround tracks. However if you play it through a normal unprocessed stereo playback system, it will sound just fine in stereo. Great stuff!

Back to the story, though. I had delivered to the projectionist a stereo, un-encoded mix, yet it was playing back in glorious 5.1 surround sound in Dolby's theater. He asked me to come back to the projection booth and have a look at the matrix (Dolby's processing box). He was right in that the decoder was taking my stereo mix and "decoding" it into a listenable, well-mixed soundtrack.

This got me thinking. Does one have to mix in surround? Is it perfectly acceptable to have a decoder convert a stereo mix into surround? The *Tulip* screening was in 2008, and these days converter boxes and software can do that for you, but I believe that one should create their own surround mix and not have an algorithm make decisions for you. There will be no placement of specific sound in the surround field. What the encoding/decoding does is take what is common to both the left and right (usually dialog) and put it in the center channel. Then it will take what is extreme left and right (usually reverbs and wide ambiences) and place them in the surround. Anything around 120 Hz and below goes to the LFE, merely acting as a sub-woofer to re-enforce low frequencies. LFE channels are meant to have specific uses, like a low-frequency impact from an explosion or a punch

DEATH HOUSE

In early 2016 I read the script for the horror film *Death House* and started writing music five months before principal shooting began. It was the fifth film that I scored and mixed for B. Harrison Smith and while he gives me excellent score direction, he also trusts my instincts. The film is about a prison break in a secret facility, and our two main characters have to go through a series of horrors before there is an unexpected end to the film. (There was the possibility of multiple sequels as the producers were eager to expand the franchise.) With all of the visual effects (VFX), combined with a thrilling plot line filled with a rollercoaster of sensational events, this was not going to be a simple soundtrack. In reading the script, I was looking for any situation where I could be quiet and give the audience a sonic rest.

The film would be shot in Philadelphia and Los Angeles. During the Philadelphia filming, I would visit the set at Holmesburg prison, an abandoned, decaying facility. I knew the production sound crew and I always like to visit the shoots and see what was going on and check if any post production audio could be recorded on set. The prison was very reverberant and Ben Wong, the production sound mixer, let me know that this was going to be a lav film, meaning that probably none of the boompole audio would be usable. I made a mental note.

The production sound crew for *Death House*

After Holmesburg prison, the crew moved to LA, where an unknown production sound team was used and there were some issues with mislabeled audio files. This was a red flag for me and made me a little anxious about how the audio was going to sound.

Once shooting wrapped, the picture was edited quickly and I received a locked picture and an OMF in mid-July. Just prior to that, I had seen edits and worked on the score so that I could have a spotting session with Harrison to discuss all aspects of the soundtrack and have some music to play for him. If

I could get the score approved, I would be ahead of schedule. Usually I am under the gun and rushing to complete a mix, but I had a sense that this would be different. During the spotting session, I realized that a lot of VFX were needed and that process takes time. When I inquired when VFX would be done, I was told that they would be complete some time in September, which meant I wanted to have the mix done at roughly the same time. In fact, VFX were not completed until the end of October but the good news was that the director approved the score with few changes. I was on my way.

PRISON SOUNDS

So *Death House* is a horror film that takes place in a prison. Everyone escapes. Shocking, I know. But, around forty minutes of the film contains the sounds of a prison break, with voices, cell doors opening, gunshots, and so on. How was I going to make that sound interesting and not a continuous din of noise? Audiences tend to stop noticing sound if it is continuous. And I had to hear the prison break from various levels inside the prison. What I did was edit together from production sound anything that sounded like prisoners escaping and fighting—and there was a lot of that. I supplemented those recordings with library sounds that were similar in nature. I placed these sounds on a timeline in Pro Tools, and actually needed only about five minutes. What I did was take this recording to the prison and play it over a loudspeaker, and then I placed six microphones around the prison and recorded the prison. While I was at it, I recorded gunshots, cell doors opening and closing, screams, and more importantly, Adrienne Barbeau's narration, which had been taped already. This narration would be the computer inside the prison, used as a voice in virtual-reality simulators, as well as announcements for the prison interior. I recorded the microphones back onto the same Pro Tools session that generated the source sounds. This way I would always be locked into the natural time delay of distant mics. I wanted the most distant mics to be placed in the surround speaker to immerse the audience in the prison. The recordings sounded fantastic: No reverb could sound that rich. And even if it could, I wanted my soundtrack to be unique. This did the trick.

Recording the prison

FOLEY—FOOTSTEPS, CLOTHING, AND SQUISHY STUFF

I have had many quality control (QC) reports come back to me because the music and effects (M&E) tracks were not fully filled in. It seems that every year, QC has become more demanding and I was determined to get it right the first time on *Death House*. In addition, this film was going to have significant foreign distribution. The director also explained to me that he wanted the overall soundtrack to feel like a video game and he directed me to listen to a few of these. I found that every sound effect

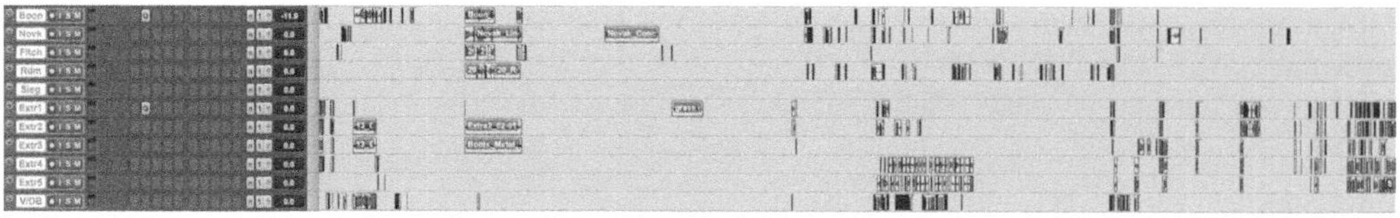

The footsteps tracks for *Death House*

that fell in the foley category was in your face and mixed very forward. With all of these issues in mind, the entire film had a separate clothing track and many separate tracks just for footsteps.

By now, you've probably guessed that this type of film is going to have blood in it. Recently I have noticed when I watch popular streaming series that the sonic character of anything that involves blood has been elevated in intensity. Every time anyone gets stabbed or shot, there is a layer of "squishyness" that is mixed pretty loud—or at least is for my personal taste. However, this is what audiences expect, and I had to comply.

I recorded the effects I needed by cooking a box of elbow macaroni, which seems to be a 'go to' item for me since I used it in *Tulip*, and taking it to my basement where there is a linoleum floor. There I set up a portable Pro Tools rig on an iMac so that I could create the sound effects to picture. With some extra water nearby in a bucket, I was able to squeeze the pasta through my fingers or smash it on the floor. I also recorded some single "squish" elements to use as hard effects just in case any were required. The director loved the sounds when he heard them.

Squishing macaroni in my basement

AMBIENCES, OR LACK THEREOF

Much like the sound effects treatment of *Death House*, Harrison wanted an ambience that reminded him of certain video games. When I listened to the games, I understood that what I was hearing were not the atmospheres of real locations, like a field or a café. What he considered ambiences were actually sounds that had been synthetically created, like a drone or a synthesized tone. Since most of *Death House*'s location was in the prison anyway, with a constant din of prisoners, I did not concern myself with real life ambient recordings but instead created long "pads" or "washes" of synthesizer sounds. What concerned me in creating these sounds is that they would not be continuous sound, which would either mask dialog, or become annoying to the audience. Also, I did not want them to have any kind of key center musically that conflict with the score and sound like someone playing a wrong note. What seemed to work was a variety of slow-moving pulses or washes that were creepy enough, yet did not impair the score or dialog.

AN EVOLVING SCORE

As Harrison and I sat down together for a number of mixing sessions, changes were made to the score even though, if you remember, in theory it had been approved months earlier. What Harrison kept asking for were elements of music from the 1980s. My first pushback was that music from that time was currently very en vogue, and I didn't want it to seem like we were chasing a trend. When it comes to music from the 80s, I was there, on the frontline, playing in bands for the whole decade. Frankly, for the past thirty years, clients have routinely asked me to remove any kind of buzzy synth from the scores that I wrote. And now, I was arguing with the director along the same lines. Harrison was right though, and this is a great example of having collaboration built on trust. I gave him what he asked for, and the revised score moved the story along while creating a hybrid concept of a dark horror score enhanced by synthesizers.

It is so important to have directors in the room with you for mixes, as you can have a real dialog and an exchange of ideas. Sending mixes back and forth, in response to notes and feedback, on the other hand, often makes the process difficult and unrewarding. In fact at times it the process seems to be endless. If the person in charge were in the room, though, I could make changes on the fly and they could react to them there and then. After a while, I begin to understand their needs, their vision for the film, and how they hear.

RELAXED MIX

I found the *Death House* mix to be a relaxing, un-stressful process. I was waiting for the visual effects people to render scenes, and as long as I was ahead of them, I could mix at a leisurely pace. This allowed me to mix frequently and for short periods of time, maybe three hours at a time—no ear fatigue on this film.

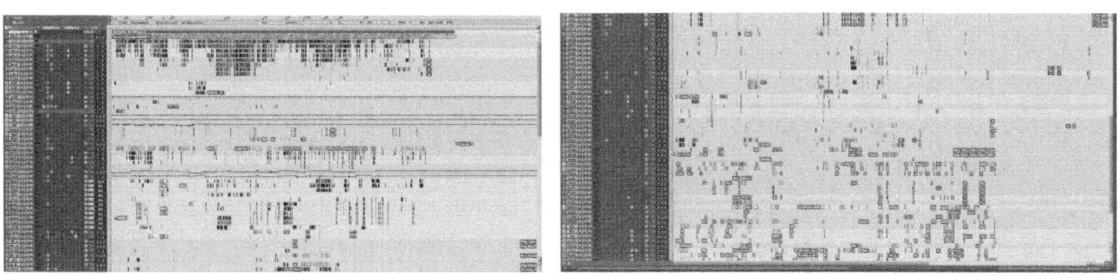

The master mix session for *Death House*

INDEX